Smart Technologies and Digital Transformation in Additive Manufacturing

The book presents the advancements and innovations in the field of additive manufacturing and its applications in diverse sectors such as medical, automotive, agriculture, aviation, and food processing. It further addresses the cutting-edge breakthroughs in intelligence systems including artificial intelligence, machine learning, augmented reality, and learning algorithms, and their impact on welding, 3D printing, and smart materials. This book:

- covers hybrid mode of troubleshooting techniques in additive manufacturing, and data-driven intelligence for enhanced production;
- highlights the implementation of intelligent systems such as artificial intelligence, machine learning, augmented reality, and learning algorithms in additive manufacturing;
- discusses intelligent methodology for design, technology, and materials optimization for wire arc additive manufacturing;
- presents additive manufacturing applications in diverse areas such as agriculture, aviation, automotive, medical, and food processing; and
- addresses the cutting-edge advancements in functionally graded materials within the realm of additive manufacturing.

The text is primarily written for senior undergraduates, graduate students, and academic researchers in fields including manufacturing engineering, mechanical engineering, aerospace engineering, automotive engineering, industrial engineering, and production engineering.

K. Sripriyan is Assistant Professor (Selection Grade) in the School of Artificial Intelligence at Amrita Vishwa Vidyapeetham, Coimbatore campus, Tamil Nadu, India.

J. Paulo Davim, is full Professor at the University of Aveiro, Portugal, holds a PhD and an MSc in Mechanical Engineering, specializing in manufacturing, materials, and industrial engineering, with a focus on machining, tribology, and surface engineering.

Smart Technologies and Digital Transformation in Additive Manufacturing

Edited by

K. Sripriyan and J. Paulo Davim

CRC Press
Taylor & Francis Group
Boca Raton London New York

CRC Press is an imprint of the
Taylor & Francis Group, an **informa** business

Designed cover image: © 2026 Shutterstock

First edition published 2026
by CRC Press
2385 NW Executive Center Drive, Suite 320, Boca Raton, FL 33431

and by CRC Press
4 Park Square, Milton Park, Abingdon, Oxon, OX14 4RN

CRC Press is an imprint of Taylor & Francis Group, LLC

ISBN: 978-1-032-94237-7 (hbk)
ISBN: 978-1-032-94238-4 (pbk)
ISBN: 978-1-003-56963-3 (ebk)

DOI: 10.1201/9781003569633

Typeset in Sabon
by Newgen Publishing UK

Contents

Preface

The rapid evolution of smart technologies and digital innovation has revolutionized the landscape of manufacturing, enabling unprecedented capabilities in design, prototyping, and production. In this context, additive manufacturing (AM) has emerged as a transformative force, breaking traditional boundaries and allowing for the fabrication of complex geometries with enhanced functionality and precision. The synergy between intelligent systems and additive processes such as welding, 3D/4D/5D/6D printing, and the integration of smart materials is driving a new era in manufacturing, one characterized by flexibility, customization, and data-driven decision-making.

This book, *Smart Technologies and Digital Transformation in Additive Manufacturing: Welding, 3D Printing, and Smart Materials*, brings together a collection of recent research contributions and technological advancements in the field of wire additive manufacturing (WAM) and related domains. The chapters investigate into how cutting-edge intelligent systems—encompassing machine learning, data analytics, automation, and control systems—are shaping the future of AM. These systems assist in optimizing parameters, predicting material behavior, ensuring quality control, and enhancing overall process efficiency.

Moreover, the book explores the pivotal role of digital transformation in bridging the gap between design and manufacturing, supported by robust frameworks for data pre-processing, feature extraction, algorithm development, statistical modeling, and validation techniques. The integration of smart materials further enhances the adaptability and functionality of printed components, offering solutions for sectors ranging from aerospace and automotive to biomedical and construction.

The insights presented herein are the result of collaborative efforts by researchers, engineers, and academicians who are pushing the frontiers of innovation. It is our belief that this compilation will serve as a valuable resource for scholars, industry professionals, and students aiming to deepen their understanding of modern manufacturing processes and their digital counterparts.

We express our gratitude to all contributors and reviewers who have helped shape the quality of this publication. It is our sincere hope that this book will inspire further exploration and development in the realm of AM and smart technologies.

Coimbatore, India
May 2025

K. Sripriyan
J. Paulo Davim

About the Editors

K. Sripriyan is Assistant Professor (Selection Grade) in the School of Artificial Intelligence at Amrita Vishwa Vidyapeetham, Coimbatore campus, Tamil Nadu, India. He has 17 years of teaching and research experience, which includes academic positions at VIT and PSG College of Technology. He completed his Master's degree in Manufacturing Engineering from AC Government College of Engineering and Technology, and his PhD in Welding Technology from Anna University, Chennai. Dr. Sripriyan has published 35 peer-reviewed papers in international and national journals and conferences. He is the inventor of four Indian granted patents, reflecting his active contributions to applied research and innovation. His research interests span additive manufacturing, moving heat source modeling, and machine learning. He has successfully completed two funded research projects sponsored by the ANRF and the Office of PSA, Department of Science and Technology (DST), Government of India. Dr. Sripriyan is a lifetime member of the Institution of Engineers (IE) and has served as editor for several academic books published by CRC Press.

J. Paulo Davim, is Full Professor at the University of Aveiro, Portugal, holds a PhD and an MSc in Mechanical Engineering, specializing in Manufacturing, Materials, and Industrial Engineering, with a focus on Machining, Tribology, and Surface Engineering. Recognized globally, he has received prestigious awards. With over 35 years of teaching and research. He has supervised numerous postdoc, PhD, and master's students and actively participated in funded research projects. As editor in chief of several international journals, he has an extensive publication record, contributing to books, chapters, and articles. Recognized as one of the World's Top 2% Scientists by Stanford University, his influence expands beyond academia, with honorary professorships in China, India, and Spain, and roles as an evaluator for prestigious bodies like the European Research Council (ERC).

Contributors

Shampa Aich
Indian Institute of Technology Kharagpur
West Bengal, India

S. Ajithkumar
KPR Institute of Engineering and Technology
Coimbatore, Tamil Nadu, India

Jaffar Ali
PSG College of Technology
Coimbatore, Tamil Nadu, India

A. Arun Negemiya
Erode Sengunthar Engineering College
Perundurai, Tamil Nadu, India

Bibhabasu Debnath
Indian Institute of Technology Kharagpur
West Bengal, India

K. Deepak Arjun
PSG College of Technology
Coimbatore, Tamil Nadu, India

N. Ganesh
KPR Institute of Engineering and Technology
Coimbatore, Tamil Nadu, India

S. Gejendhiran
Madanapalle Institute of Technology and Science
Madanapalle, Andhra Pradesh, India

Ramkumar Gunasekaran
Coimbatore Institute of Technology
Coimbatore, Tamil Nadu, India

Gouri Vignesh Jawalkar
Indian Institute of Technology Kharagpur
West Bengal, India

G. Kalusuraman
Kalasalingam Academy of Research and Education
Krishnankoil, Tamil Nadu, India

K. Karthigadevi
Kalasalingam Academy of Research and Education
Krishnankoil, Tamil Nadu, India

M. Karthigha
PSG Institute of Technology and Applied Research
Neelambur, Tamil Nadu, India

R. Kathiravan
PSNA College of Engineering and Technology
Dindigul, Tamil Nadu, India

Sibani Mahapatra
Indian Institute of Technology Kharagpur
West Bengal, India

S. Manikandan
Erode Sengunthar Engineering College
Perundurai, Tamil Nadu, India

P. B. Mithiran
PSG College of Technology
Coimbatore, Tamil Nadu, India

K. S. Mohanraj
Erode Sengunthar Engineering College
Perundurai, Tamil Nadu, India

Rupshree Ozah
Madanapalle Institute of Technology and Science
Madanapalle, Andhra Pradesh, India

R. Prithivirajan
Madanapalle Institute of Technology and Science
Madanapalle, Andhra Pradesh, India

R. S. Raghul
PSG College of Technology
Coimbatore, Tamil Nadu, India

Rajesh Ranganathan
Coimbatore Institute of Technology
Coimbatore, Tamil Nadu, India

Y. Robinson
Erode Sengunthar Engineering College
Perundurai, Tamil Nadu, India

Abhishek Saha
Indian Institute of Technology Kharagpur
West Bengal, India

N. Saravanan
Erode Sengunthar Engineering College
Perundurai, Tamil Nadu, India

S. Sivasankari
Erode Sengunthar Engineering College
Perundurai, Tamil Nadu, India

G. Srinivasagan
Rajapalayam Rajus College
Rajapalayam, Tamil Nadu, India

K. Sripriyan
School of Artificial Intelligence, Amrita School of Engineering, Amrita Vishwa Vidyapeetham
Coimbatore, Tamil Nadu, India

S. Suresh
Erode Sengunthar Engineering College
Perundurai, Tamil Nadu, India

S. Syath Abuthakeer
PSG College of Technology
Coimbatore, Tamil Nadu, India

K. Vignesh
PSNA College of Engineering and Technology
Dindigul, Tamil Nadu, India

Chapter 1

Introduction to wire additive manufacturing and regulatory framework

R. Kathiravan and K. Vignesh

1.1 INTRODUCTION

Three dimensions are typically used in traditional additive manufacturing, where the item is constructed by moving the print head or nozzle in the X, Y, and Z axes. In the context of wire arc additive manufacturing (WAAM), however, the terms four- (4D), five- (5D), and six-dimensional (6D) refer to expansions of the conventional 3D process that offer greater control and movement capabilities, allowing for the creation of more complicated parts and new features presented in Table 1.1.

Six degrees of freedom (6DoF) is a term used to describe the freedom of movement of a rigid body in 3D space. It encompasses the ability to move in three translational directions (up/down, left/right, forward/backward) and to rotate about three perpendicular axes (pitch, yaw, and roll). The freedom of motion of a rigid body in three dimensions is referred to as 6DoF, as shown in Figure 1.1. It includes the capacity to spin around three perpendicular axes (pitch, yaw, and roll) and travel in three translational directions (up/down, left/right, and forward/backward). In more detail, the 6DoF are defined as follows (Ge et al., 2014):

X-axis translation: Shifting to the left or right.
Y-axis translation: Moving up or down.
Z-axis translation: Moving forward or backward.
Rotation: tilting upward or downward about the X-axis (pitch).
Rotation: Turning left or right about the Y-axis (Yaw).
Rotation about the Z-axis (roll): Tilting side to side.

The idea of 6DoF is essential in several domains, such as computer graphics, robotics, virtual reality, and aircraft. For example, it enables the precise control of robotic arms and vehicles in robotics, allowing them to carry out complex tasks and negotiate challenging settings. 6DoF in virtual reality improves the user experience by enabling organic movement and interaction in a virtual environment. It is a key idea in contemporary technology

DOI: 10.1201/9781003569633-1

Table 1.1 Basic printing techniques characteristics, limitations

Degree of freedom	*Basic characteristics*	*Limitations*	*Equipment requirement*
3D	Layer-by-layer development in the vertical axis	Lack of adaptability regarding the layer orientation and the product lacking intellect	3D printers
4D	Creation of smart material products	Restrictions on the orientation of the layer Strongly directed and just slightly programmable intelligences	3D printer raw material with intelligence
5D	Layering in all directions, less material and process time.	Product with no intelligence, higher setup cost. Limited for extreme multi-directional geometries.	Robotics arm
6D	Layering in all directions		
Less time spent on materials and processes. Development of Smart Material Products with adaptable and modifiable programmability	Additional calibration and the need for standardization. Higher equipment costs, complex programming, and operational challenges.	Robotics arm raw material with intelligence	

and engineering; 6DoF must be understood and applied to build systems that need precise spatial awareness and manipulation. Understanding and implementing 6DoF is essential for creating systems that require accurate spatial awareness and manipulation, making it a fundamental principle in modern technology and engineering.

1.2 4D WAAM

Time as the fourth dimension: In the context of WAAM, the "4D" refers to adding a time-dependent component to the standard 3D process. It does not just involve spatial movement (X, Y, and Z), as well as the way the geometry evolves or changes over time. By adding the element of time, 4D printing enables printed items to gradually alter their shape or characteristics in reaction to outside stimuli (such as pH, moisture, or temperature). The development of structures that can adjust to their surroundings is made possible by this dynamic flexibility (El-Sayegh et al., 2020).

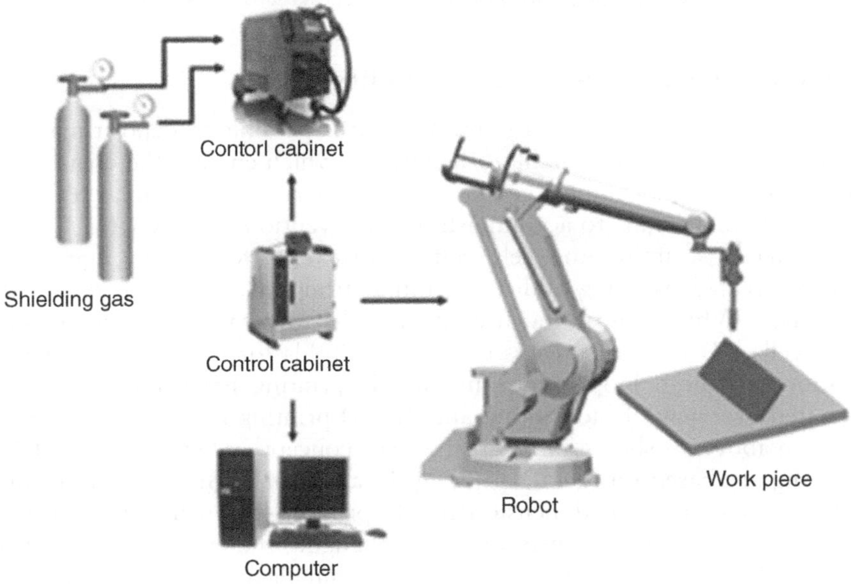

Figure 1.1 Freedom of motion.

1.2.1 Adaptive manufacturing

The capacity to alter the welding head's path or the rate of deposition over time in response to changes in the material's characteristics, the geometry of the component, or even external variables like temperature is known as adaptive manufacturing. The manufacturing processes' adaptability is crucial in applications such as producing parts with variable qualities or features that alter throughout production (Alam et al., 2023).

1.2.2 Shape memory alloys

Shape-memory alloys are a notable example of a material that uses the 4D concept. Parts made with WAAM can alter shape or structure in response to external stimuli, such as heat or mechanical stress. Research in this area is ongoing.

1.3 5D WAAM

Extra control over the deposition process: Adding a rotational axis to the system is sometimes referred to as the "fifth dimension" in architecture. This indicates that the print head or welding torch can revolve around an axis, usually known as the A (tilt) and B (rotation) axes, in addition to moving in the X, Y, and Z directions.

1.3.1 Increased geometric flexibility

By including these rotational movements, it is possible to deposit material more accurately in challenging orientations, which enables the creation of more intricate and complex geometries, like overhangs or angular features, which are challenging to accomplish with conventional 3D WAAM.

Advanced scaffolds and cell-scaffold structures for tissue engineering can be created by using additive manufacturing, also referred to as "3D printing." While bioprinting uses living cells to create structures loaded with cells, 4D printing creates dynamic scaffolds for human tissue regeneration. But both bioprinting and 3D/4D printing have drawbacks. An updated overview of bioprinting and 3D/4D printing in tissue engineering is given above, as shown in Figure 1.2. The concept and explanation of 5D printing are based on 3D/4D printing. In addition to 3D space and time, information—the fifth dimension in 5D printing—is included in printed structures and can be transferred later, changing the environment of 5D-printed things. 5D printing creates active or intelligent objects that interact with their surroundings and have a beneficial impact, in contrast to 3D/4D printing, which creates inert or inactive items, as shown in Figure 1.3 (Lai & Wang, 2023).

1.3.2 Better surface quality and accuracy

By enabling more control over the nozzle orientation during deposition, rotational axes lower the possibility of flaws and enhance the finished part's overall surface polish. In addition, it makes multi-pass welding more accurate for improved material bonding (Sripriyan et al., 2022).

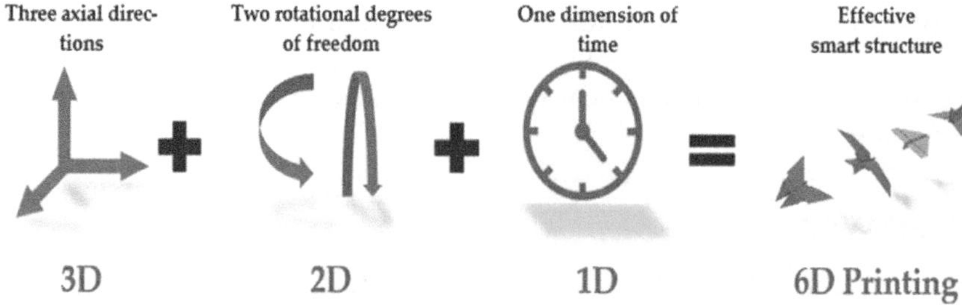

Figure 1.2 Better surface quality and accuracy.

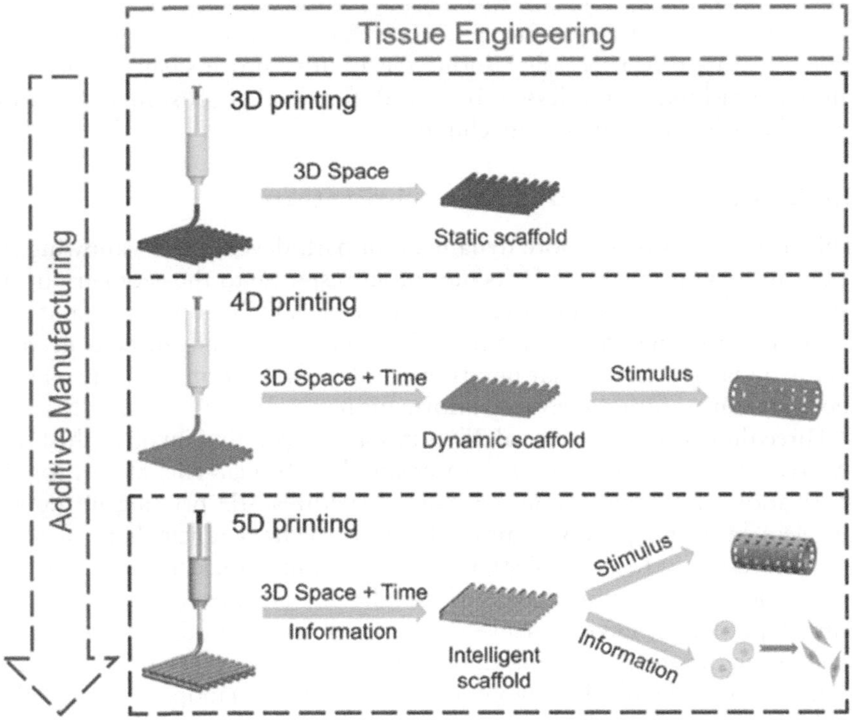

Figure 1.3 In situ information delivery.

1.3.3 In situ information delivery

In 5D printing, the embedded information can be delivered in situ during the application of the printed structures. This means that the information can be released at the right time and place to influence cellular behavior or environmental conditions, enhancing the functionality of the printed object.

1.4 6D ADDITIVE MANUFACTURING USING WIRE ARC

Full 6-axis control: The printing head or robotic arm in the 6D WAAM process has full 6-axis control, which enables it to move in all three rotating axes (A, B, and C) and translational axes (X, Y, and Z). This enables even more precise control over the part orientation and deposition path during manufacture (Srivastava et al., 2019).

1.4.1 Increased accuracy and capability

6D control enables the system to process even more intricate shapes, carry out more intricate post-processing during the additive process (such as accurate surface finishing), and lessen the need for further post-manufacturing procedures like polishing or machining.

1.4.2 Automated tool path optimization

This feature allows for more dynamic tool path development, allowing the robot to dynamically modify its motion in response to the geometry of the part it produces. Consequently, manufacturing is highly optimized and subject to fewer geometrical constraints. 5D printing is an advanced concept in additive manufacturing that builds upon 3D and 4D printing principles by incorporating a fifth dimension: information.

Three dimensions traditional 3D printing, 5D printing involves the layer-by-layer deposition of materials to create 3D objects. This allows for the fabrication of complex geometries and structures. 4D printing introduces the element of time, allowing printed objects to change their shape or properties in response to external stimuli (such as temperature, moisture, or pH) over time. This dynamic capability enables the creation of structures that can adapt to their environment. The defining feature of 5D printing is the integration of information into the printed structures. This information can be in the form of embedded biomolecules, functional nanoparticles, or other entities that can trigger changes in the environment or the printed object itself upon release. The information can guide the behavior of the printed structure or influence the surrounding environment, leading to desired outcomes (Rathee & Srivastava, 2023).

1.4.3 Bending deformation in 4D and 6D printing

The difference in bending deformation between 4D and 6D printing lies primarily in the capabilities, and mechanisms of each technology involve the creation of 3D-printed objects that can change their shape or properties over time in response to external stimuli (such as temperature, moisture, or pH). The bending deformation in 4D printing is typically limited to the changes that occur due to these stimuli, which may require additional processing or support materials to achieve the desired shape change. The printed structures are often designed to respond in a specific way, but the range of deformation is constrained by the materials and the initial design.

As shown in Figure 1.4, the difference between the geometrical shape of the effect on 4D and 6D printing is represented as horizontal layering in 4D shape of the object, as well as inclined v-type layering deposition. In contrast, 6D printing builds upon the principles of 5D printing by incorporating smart materials that can respond more dynamically to external stimuli. This

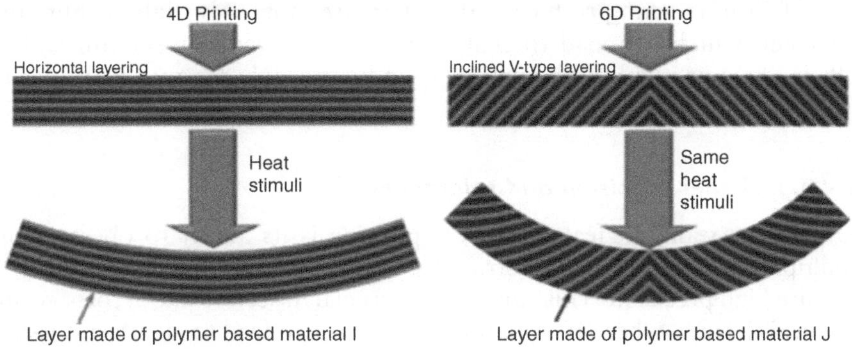

Figure 1.4 4D and 6D printing.

technology allows for the production of objects that can exhibit more complex bending and deformation behaviors (Sripriyan et al., 2023). The introduction of additional rotational axes in 6D printing enables the creation of structures that can bend and deform in multiple directions and angles, resulting in greater flexibility and adaptability compared to 4D-printed objects. This capability allows for more sophisticated designs that can better respond to environmental changes.

In summary, while 4D printing allows for shape changes in response to stimuli, 6D printing enhances this capability by enabling more complex and versatile bending deformations through the use of advanced materials and additional dimensional control. This results in improved performance and functionality of the printed objects (Ge et al., 2014).

1.4.4 Advantages of 6D printing

1.4.4.1 Enhanced structural intelligence

6D printing enables the development of smart materials that can adapt their shape or properties in response to external stimuli, such as temperature, light, or magnetic fields. This capability allows for the creation of structures that can actively respond to their environment, making them more versatile and functional in various applications. For instance, a 6D-printed object could change its configuration to optimize performance under different conditions, which is a significant advancement over traditional static materials (Youssef et al., 2024).

1.4.4.2 Improved manufacturing efficiency

The 6D printing process is designed to streamline production by reducing the amount of material wasted and speeding up the manufacturing timeline. By using advanced techniques and multi-axis printing, manufacturers can produce parts that are not only lighter but also stronger, which can lead to cost savings and faster delivery times. This efficiency is particularly beneficial in industries where rapid prototyping and production are essential.

1.4.4.3 Higher precision and tolerances

One of the standout features of 6D printing is its ability to operate with multiple axes, allowing for intricate designs and complex geometries that require high precision. This capability is crucial in sectors like aerospace and medical devices, where even minor deviations can lead to significant issues. The ability to produce components with tighter tolerances ensures that they meet stringent industry standards and perform reliably.

1.4.4.4 Broader application potential

The anticipated quick adoption of 6D printing across various industries highlights its versatility and potential impact. Fields such as aerospace, medical devices, and electronics stand to benefit significantly from the technology's superior accuracy and functional capabilities. For example, in the medical field, 6D printing could lead to the creation of customized implants that better fit patients' anatomical needs.

1.4.4.5 Material optimization

6D printing not only enhances the strength and quality of the final products but also minimizes the time required for processing. This optimization means that manufacturers can produce higher-quality materials that perform better under stress or in specific conditions, ultimately leading to more durable and effective products.

In summary, the introduction of 4D, 5D, and 6D capabilities in WAAM represents a significant leap toward more flexible, precise, and efficient manufacturing, especially for complex, high-performance parts. These capabilities allow manufacturers to move beyond the limitations of traditional 3D printing and unlock new possibilities for production across various industries (Nida et al., 2022).

1.5 EMERGING APPLICATIONS OF 5D AND 6D PRINTING IN THE FOOD INDUSTRY

5D printing can be used to create intricate food items such as ball-shaped confectionery, concave-shaped snacks, and other complex geometries that are difficult to achieve with traditional 3D printing methods. This capability allows for innovative product designs that enhance consumer appeal, as shown in Figure 1.5 (Nida et al., 2022).

1.5.1 Food packaging

The technology can be applied to develop complex-shaped food packaging materials, such as curved boxes and tableware (cups and bowls). This is particularly advantageous as it eliminates the need for additional support materials, making the packaging process more efficient.

1.5.2 Smart food products

6D printing introduces the use of smart materials that can respond to external stimuli, such as temperature or pH changes. This can lead to the development of food products that can change their properties (e.g., flavor, texture) in response to environmental conditions, enhancing the consumer experience. Both 5D and 6D printing can be used to create food products with tailored nutritional profiles. For instance, the incorporation of specific nutrients or functional ingredients can be achieved through precise printing techniques, allowing for personalized nutrition solutions. The efficiency of 5D printing in terms of material usage and the potential for reduced waste in food manufacturing processes can contribute to more sustainable practices

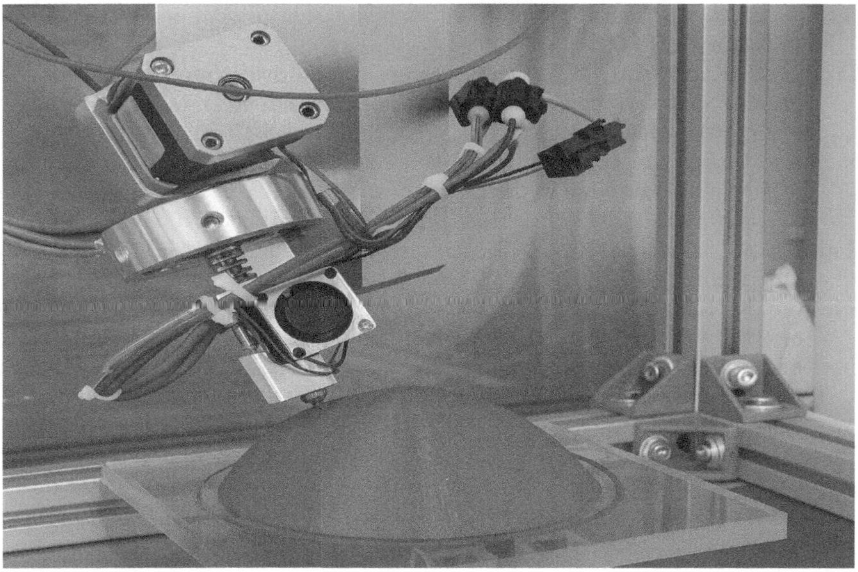

Figure 1.5 Food packaging.

in the food industry. This aligns with the growing demand for environmentally friendly production methods.

These emerging applications highlight the potential of 5D and 6D printing to revolutionize food manufacturing, offering innovative solutions that cater to modern consumer demands and sustainability goals.

1.6 CONCLUSION

Cutting-edge developments in the area of additive manufacturing are represented by 4D, 5D, and 6D printing. Dynamic, self-transforming materials are introduced by 4D printing, optimal material deposition is used by 5D printing to increase structural strength, and real-time intelligence and multifunctionality are integrated by 6D printing. These technologies offer more intelligent, robust, and flexible answers to today's problems, with the potential to revolutionize industries including healthcare, aircraft, and energy.

REFERENCES

Alam, Z., Iqbal, F., & Khan, D. A. (2023). *Post-Processing Techniques for Additive Manufacturing*. CRC Press. https://doi.org/10.1201/9781003288619

El-Sayegh, S., Romdhane, L., & Manjikian, S. (2020). A critical review of 3D printing in construction: benefits, challenges, and risks. *Archives of Civil and Mechanical Engineering*, 20(2), 34. https://doi.org/10.1007/s43452-020-00038-w

Ge, Q., Dunn, C. K., Qi, H. J., & Dunn, M. L. (2014). Active origami by 4D printing. *Smart Materials and Structures*, 23(9), 094007. https://doi.org/10.1088/0964-1726/23/9/094007

Lai, J., & Wang, M. (2023). Developments of additive manufacturing and 5D printing in tissue engineering. *Journal of Materials Research*, 38(21), 4692–4725. https://doi.org/10.1557/s43578-023-01193-5

Nida, S., Moses, J. A., & Anandharamakrishnan, C. (2022). Emerging applications of 5D and 6D printing in the food industry. *Journal of Agriculture and Food Research*, 10, 100392. https://doi.org/10.1016/j.jafr.2022.100392

Rathee, S., & Srivastava, M. (2023). *Wire Arc Additive Manufacturing*. CRC Press. https://doi.org/10.1201/9781003363415

Sripriyan, K., Ramu, M., & Thyla, P. (2023) Characterization of dissimilar welding of AISI 409 ferritic to AISI 316l austenitic stainless steel using AISI 308 filler wire under different temperature conditions, in *Advances in Additive Manufacturing and Metal Joining: Proceedings of AIMTDR 2021* (pp. 309–321). Springer.

Sripriyan, K., Ramu, M., & Thyla, P., Anantharuban K., & Karthigha M (2022) Characteristic of weld bead using flat wire electrode in GMAW inline during the process: an experimental and numerical analysis. *International Journal of*

Pressure Vessels and Piping, 196(January). https://doi.org/10.21203/rs.3.rs-847159/v1
Srivastava, M., Rathee, S., Maheshwari, S., & Kundra, T. K. (2019). *Additive Manufacturing*. CRC Press. https://doi.org/10.1201/9781351049382
Youssef, H., El-Hofy, H., & Ahmed, M. (2024). *Fundamentals of Additive Manufacturing*. CRC Press. https://doi.org/10.1201/9781003451440.

Chapter 2

Material processes and AI advancements in additive manufacturing

Gouri Vignesh Jawalkar, Abhishek Saha, Bibhabasu Debnath, Sibani Mahapatra, and Shampa Aich

2.1 INTRODUCTION

Additive manufacturing (AM) is depositing material, layer by layer, from a three-dimensional (3D) computer-aided design (CAD) model. Compared to the traditional subtractive manufacturing techniques, they typically have lower material wastages and higher efficiencies when applied to various materials, including polymers, metals, ceramics, glass, composites, and hybrids, with more considerable complexities in geometry [1, 2]. Lower metal wastage becomes essential for maintaining the overall economy of the product, especially in the case of complex material systems like superalloys, oxide dispersion-strengthened (ODS) steels, high entropy alloys, etc. Thus, AM is widely explored in nuclear [3, 4] and aerospace sectors [5, 6]. Additionally, the complexity achieved by AM processes is of great advantage for the aerospace sector because it allows us to maintain a high fly-to-buy ratio with enhanced mechanical and thermal performance and reduced system mass, which is not possible with other manufacturing methods.

Recently, the first 3D-printed rocket engine part was launched by SpaceX in its Falcon 9 series, which proved to have better strength, fracture resistance, and ductility than a part made with traditional casting, as well as lower variability in material properties [7]. The benefits of AM have also been explored beyond the boundaries of Earth to provide in-space manufacturing at the International Space Station, expanding the horizons of scientific exploration [8]. The large structures, however, are usually built in parts and assembled later to form the final product, which marks one of their disadvantages.

Beyond significant structural and functional components, AM is also explored in devices like micro-electromechanical systems [9] and biomedical industries [10, 11]. The flexibility of AM for these advanced material systems is also explored at microscales for electronics and semiconductor applications. For instance, Norsk Titanium and Hittech Group developed a Ti6Al4V carrier tray for advanced semiconductor materials lithography using a wire directed energy deposition (DED)-based AM technique, which

DOI: 10.1201/9781003569633-2

reduced mass wastage by almost 60% and energy consumption savings by nearly 45% [12]. Multiple efforts have recently been made to integrate additive and subtractive manufacturing [13].

Indeed, AM is rapidly evolving and has been an emerging field in the last few years and will be in the future. However, it also faces challenges like inconsistent quality production, limited process control, and the optimization of manufacturing parameters [14]. They can significantly affect the reliability and scalability of AM processes. This can severely impact critical applications like aerospace, medical devices, and automotive parts. Here, machine learning (ML) and artificial intelligence (AI) come into the picture, analyzing large datasets and optimizing the process to a greater extent. This chapter will show how ML integration has helped the entire AM process, including pre-processing, in-processing, and post-processing. Then, look at the recent developments that ML has made in AM and look forward to possible and ongoing future developments.

This chapter thus explores various material processes under wire-AM technologies and the integration of AI at each manufacturing step.

2.2 MATERIAL PROCESSES

Despite wide variations in the technologies used in AM, the AM process can usually be described in the fundamental stages of pre-processing, manufacturing, and post-processing. Designing is one of the fundamental pre-processing steps, where a 3D model of the final product is made using a CAD design package. This model is then usually exported as an STL file, which is essential to create a surface mesh out of the 3D model, making it independent of the type of package used. This exported 3D model is now sliced using software like Ultimaker to provide a toolpath or the "G-Code" for the 3D printer. This is entirely a digital phase for 3D printing, and it depends hugely on the type of printer used; see details in Figure 2.1.

Another part of pre-processing is material selection. This depends upon the type of application, and various material properties like thermal, mechanical, optical, electrical, and magnetic properties are considered before finalizing it. Different AM processes have different material printing possibilities. AM offers many materials, including ceramics, polymers, metals/alloys, composites, and hybrids. Still, based on the type of material used, the ideal process is to be selected. Once again, the chosen material must be fed with wire or powders. The material fabrication technique for the feed material and each of its parameters is thus also a part of this pre-processing stage.

Next is the fabrication stage, where the type of printing technology is selected. Based on the energy source, feed materials, feed type, technology involved, etc., there can be several classifications of AM technologies. Figure 2.2 provides a clear understanding of the standard classification of

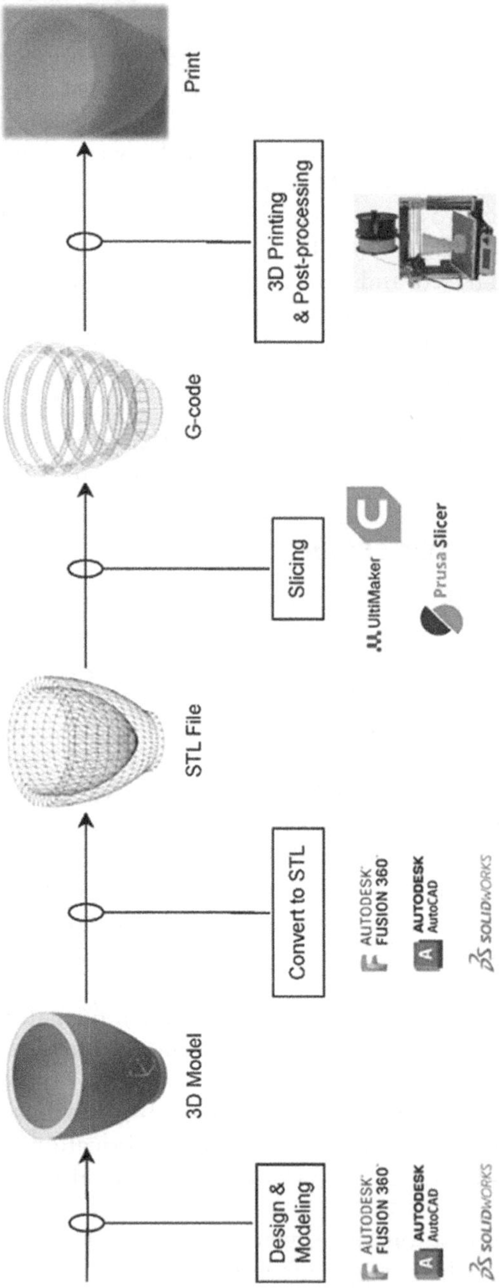

Figure 2.1 Printer types.

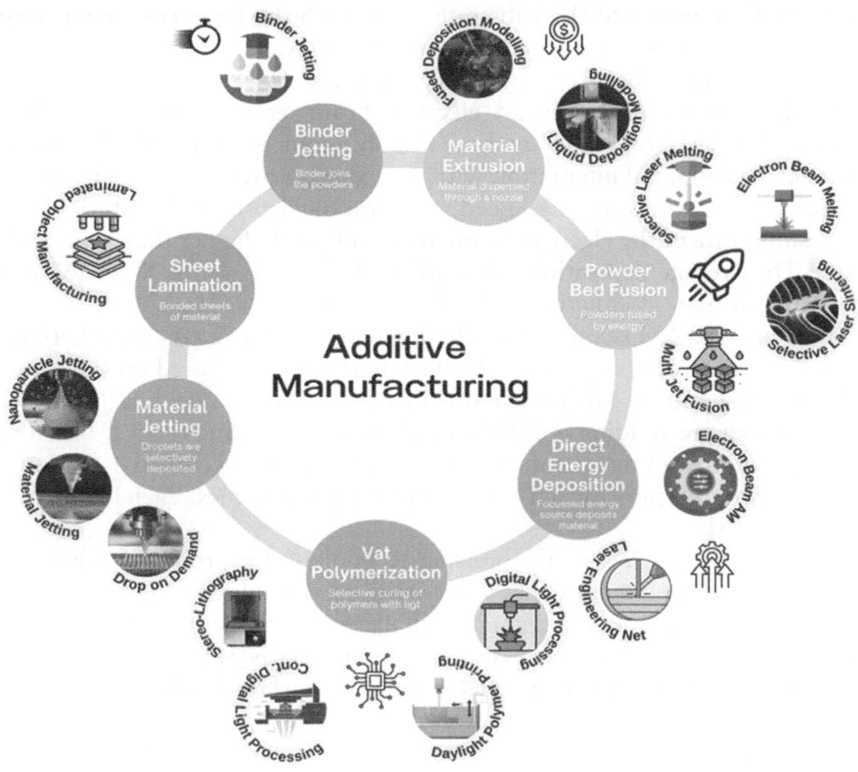

Figure 2.2 AM techniques.

AM techniques. The feed material types used for these AM processes are usually powders or wires. Powder-based AM technologies are widely explored compared to wire-based technologies, owing to the benefits of higher geometrical accuracy and compatibility with complex material systems like superalloys. The typical layer thickness in powder-feed/-bed technology is 20–100 μm, and the completed components can achieve a dimensional accuracy of ±0.05 mm and a surface roughness of 9–16 μm. However, the deposition rates of powder-based AM processes are usually longer than those of wire-based systems. The wire-based technologies can be classified based on the deposition source: laser-based, arc welding-based, and electron beam-based. The material usage in the case of these processes is usually 100%, as all the material gets deposited to form the component with minor wastages, making it more environmentally sustainable and economical.

The wire and laser AM (WLAM) laser is used as the energy source to create a melt pool on the substrate material, onto which the metal wire is fed and melted, forming a metallurgical bond with the substrate. The laser head is often moved to the substrate along with the wire, creating a bead. The process can be further fed from the front, back, or side based on the relative motion of the laser concerning the bead [15].

Electron beam AM uses an electron beam as the energy deposition source. The electron beam method usually requires an ultrahigh vacuum that ensures minimal interaction with the external environment. The energy density, efficiency, scanning speed, penetration depth, and material layer deposition rate of the electron beam are significantly higher than that of the laser. The average deposition rate and layer thickness in the WEBAM process are less than 19.8 kg/h and 3000 μm, respectively [16, 17].

Wire and arc AM is another method, where an electric arc forms between metal and electrode, depositing the wire-feed material. Based on the source of the arc, it can be further classified as gas metal arc welding (GMAW), gas tungsten arc welding (GTAW), or plasma arc welding (PAW). Based on the type of arc and metal contact, they can be further classified as globular, short-circuiting, spray, or pulsed spray transfer techniques. Each has its pros and cons [18].

While wire-feed AM is a promising technology for producing more significant aerospace components with moderate complexity, such as flanges or stiffened panels, a few technical challenges remain, including residual stress and distortion from excessive heat input, relatively poor part accuracy caused by the "stair stepping" effect, and poor surface finish.

The choice of AM techniques depends on feed type, energy source, and material properties. Table 2.1 summarizes the most common AM techniques, their key features, advantages, challenges, and relevant industries.

Oftentimes, the AM components require a few post-processing steps. This is essential to improve the product's mechanical properties and provide a surface finish for the application. Thus, techniques like sanding, polishing, and heat treatment are commonly used.

2.3 CHALLENGES AND SMART SOLUTIONS

2.3.1 Integration of AI in AM

ML integration in AM has made a meaningful change. It is essential to address the challenges AM faces by enabling the analysis of large, complex datasets generated during AM processes. ML can be used to optimize process parameters, detect defects in real time, and predict the performance of the printed parts, which thereby helps improve overall efficiency and quality. This integration of ML is undoubtedly the key to advancing AM toward more intelligent, automated, and optimized manufacturing systems. This chapter will explore how ML integration has improved the quality and

Table 2.1 Wire and powder AM techniques

AM technology	Feed type	Energy source	materials	Key features	Advantages	Challenges	Relevant Industries
Selective laser sintering (SLS)	Powder	Laser	Polymers, ceramics, hybrids, and composites	Use a high-powered laser to selectively sinter powders.	Suitable for polymers and ceramics and enables complex geometries with minimal support structures.	Surface finish may require post-processing; limited to specific materials.	Aerospace, biomedical (prosthetics, surgical tools), consumer goods (prototypes), electronics (housings).
Direct metal laser sintering (DMLS)	Powder	Laser	Metals/alloys	Melts metal powder layer by layer to produce fully dense parts.	High-dimensional accuracy; good mechanical properties suitable for metal superalloys.	High operational cost; slower build speeds compared to wire-based methods.	Aerospace (turbine components), biomedical (implants), automotive (engine parts), tooling (injection molds).
Electron beam melting (EBM)	Powder	Electron beam	Metals/alloys	Operates in a vacuum, using an electron beam to melt and fuse powdered material into fully dense components.	High-energy density; reduced oxidation risk; suitable for high-performance materials like titanium alloys.	Require vacuum environment; rougher surface finish than laser-based methods.	Aerospace (engine components), biomedical (orthopedic implants), defense (missile components).

(*continued*)

Table 2.1 (Cont.)

AM technology	*Feed type*	*Energy source*	*materials*	*Key features*	*Advantages*	*Challenges*	*Relevant Industries*
Binder jetting	Powder	Binder adhesive	Ceramics, metals/alloys, polymers, hybrids	Deposits a liquid binder onto a powdered bed, layer by layer, to bind particles and create a part.	Fast production speeds; cost-effective; supports various materials, including ceramics and metals.	Requires post-processing for full strength (e.g., sintering or infiltration).	Consumer goods (decorative items), automotive (lightweight parts), industrial (tooling), biomedical (scaffolds).
WLAM	Wire	Laser	Metals/alloys	Use a laser to create a melt pool into which a metal wire is fed and melted to form a part.	High deposition rates; minimal material waste; good metallurgical bonding.	Sensitive to heat input; stair-stepping effect may affect accuracy.	Aerospace (frames, panels), automotive (chassis components), shipbuilding (hulls).
Wire and arc additive manufacturing (WAAM)	Wire	Electric arc (GMAW, GTAW, PAW)	Metals/alloys	Employs an electric arc to melt wire feedstock deposited layer by layer to form components.	High material utilization; suitable for large aerospace and structural components.	Residual stresses and distortion due to heat; relatively poor surface finish.	Aerospace (significant structural components), oil and gas (pipelines), marine (propellers, hull components).

Electron beam additive manufacturing (EBAM)	Wire	Electron beam	Metals/alloys	It uses an electron beam in a vacuum to melt wire feedstock and create dense metal parts	High deposition rates; excellent material efficiency; suitable for large structural components.	Requires a vacuum, limited material selection, and expensive equipment.	Aerospace (rockets, frames), defense (armor, weaponry), marine (propulsion systems), energy (wind turbine parts).

implementation of AM across its various stages and investigate the recent developments regarding this integration. Additionally, anticipate the potential of ML to develop more optimized and effective solutions for AM.

2.3.2 AI-driven innovations across AM stages

The AM process can be broadly divided into three phases: pre-processing, in-processing, and post-processing. This section will explore how ML integration has helped in all three sections; see details in Figure 2.3.

2.3.2.1 *ML in the pre-processing phase of AM*

Pre-processing in AM involves designing the required structure using 3D CAD tools. Design for additive manufacturing (DfAM) is a specialization within design for manufacturing and assembly (DfMA) that leverages the

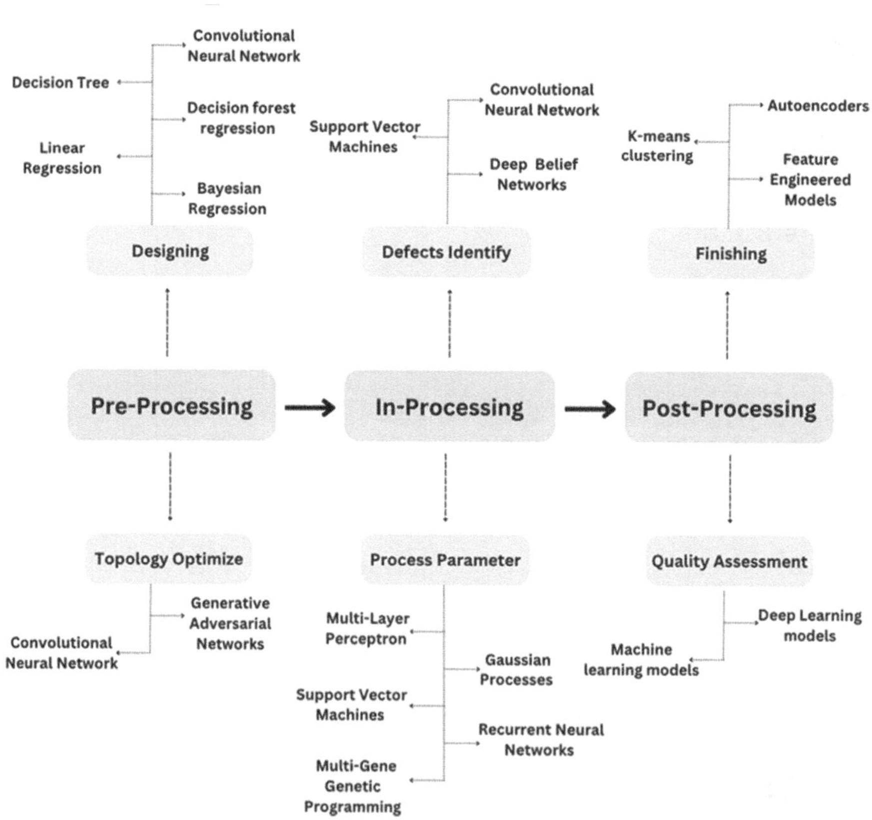

Figure 2.3 ML integration.

unique capabilities of AM [19, 20]. It focuses on designing with the AM process to optimize production efficiency and product performance. DfAM optimizes design complexity and material usage while accounting for specific production limitations. Hierarchical clustering, as employed by Yao et al. [21], aids less experienced designers by recommending feasible design features such as optimal placement of support structures. Zhou et al. [22] also developed an automated decision-making system to improve efficiency and reduce costs. This system utilizes ML algorithms like linear regression, Bayesian regression, neural networks, and decision trees (DTs). Zhou et al. [22] advanced this approach by implementing DNN and convolutional neural network (CNN) models to analyze relationships between design variables and performance.

While ML-driven design tools enhance pre-processing, topology optimization (TO) further refines these designs by minimizing material use and maximizing performance. This transition to TO is crucial for creating efficient, high-strength structures in AM. AM allows the creation of complex geometries; topological optimization has become especially important in optimizing these complex structures. TO [23] is a mathematical method that optimizes material layout within a given design space for a given set of loads, boundary conditions, and constraints to maximize the system's performance. It helps determine the optimal geometry that maximizes strength and stiffness while using the least amount of material possible. While topological optimization is essential, a typical topological orientation is computationally demanding, as it requires numerous iterations of design and prototyping. So, intense models, especially neural networks, would face many challenges in the training process. Once the ML models are well-trained, favorable designs can be retrieved quickly without starting from nothing. This enables the ML-centric method to serve as a complementary alternative to the conventional TO approach [24]. Various works and advanced models have been trained to improve the topological optimization process, like generative adversarial networks (GANs) [25] and advanced CNN architectures. However, the integration of ML in AM-based TO is still in its initial stages, and more research is needed till it becomes an integral part of the AM design process.

ML is widely used in DfAM to predict critical parameters like geometric accuracy and surface finish. Table 2.2 highlights case studies showcasing how ML techniques enhance DfAM processes.

2.3.2.2 ML in in-processing phase of AM

The in-processing phase is where the required part's fabrication process operation occurs. This step is responsible for directly influencing the quality of the manufactured part and will also determine the precision and performance of the part. These steps are essential to control the parameters so that

Table 2.2 A case study of ML usage in DfAM [22]

AM technique	*Objectives*	*ML techniques*
Wire arc DED	Predict the width and height of the printed bead	Neural network
Laser DED	Predict the depositing height of printed parts	BPNN, LS-SVM
Material extrusion	Predict geometric accuracy of printed parts	Self-organizing map, K-means unsupervised clustering
Powder bed fusion	Predict dimensional accuracy of printed parts	MLP, CNN
Material extrusion	Predict surface roughness of printed parts in different process parameters	Random forests, AdaBoost, SVM
Material extrusion	Predict printable bridge length	BPNN

the part meets its desired specifications. The in-processing phase becomes essential for AM, as it follows the layer-by-layer fabrication approach. The challenges faced using the traditional approach led researchers to work on ML-integrated processes, which have worked predominantly, as ML can analyze large datasets, capture complex, non-linear relationships, and predict process outcomes based on parameter variations. At the mesoscale, ML techniques such as multi-layer perceptron (MLP) and Gaussian processes (GP) [26] are used to predict melt pool geometry [27] porosity using reverse engineering. At the macroscale, ML models like adaptive-network-based fuzzy inference systems (ANFISs) [24] and support vector machines (SVMs) optimize mechanical properties and surface quality. Other advanced approaches like recurrent neural networks (RNN) and multi-gene genetic programming (MGGP) are also used to analyze complex tasks like thermal history prediction. ML's unique ability to capture non-linear relationships makes it a powerful tool that enhances the efficiency of the overall AM process, reduces costs, and improves quality in AM processes.

AM processes face several processing defects. These include cracks, porosity, lack of fusion, and surface roughness, often caused by the layer-wise material deposition approach. Traditional monitoring methods involve complex steps, so ML offers innovative defect detection and process optimization solutions using acoustic and optical signals. Acoustic monitoring efficiently captures high-resolution data, offering a cost-effective approach to defect analysis. They provide high temporal resolution. For powder bed fusion (PBF) [28], signals generated by plasma density variations during heating variations are processed using ML models and deep belief networks (DBN) for defect classification. In fused deposition modeling (FDM), acoustic patterns are analyzed using clustering techniques and hidden semi-Markov models to detect anomalies, demonstrating faster data processing and reduced complexity. High-speed infrared cameras monitor AM defects

Table 2.3 A Case Study of ML usage in the process of AM

AM technique	*Objectives*	*ML techniques*
SLM	To predict porosity and microhardness	MLP
DED	To estimate the process parameters required to obtain a specified melt pool geometry	MLP
DED	To predict the high-dimensional thermal history of complex parts during printing	RNN
FDM	To optimize process parameters and improve viscoelastic responses	MLP
SLM	Detect anomalies of melt tracks	Support vector machines, convolutional neural networks
DED	Detect the location and size of pores	Self-organizing maps
SLM	Detect and locate anomalies	Support vector machines

SLM, selective laser melting.

by capturing optical signatures, providing real-time insights as they can capture detailed plume, spatter, and melt pool profiles [29]. These optical inputs offer information about defect formation and act as input features for the deep learning models (Table 2.3).

2.3.2.3 ML in the post-processing phase of AM

The post-processing phase refers to the final set of operations performed before the fabricated part is ready for production in its respective industry. These processes address surface roughness, internal defects, structural inconsistencies, and other finishing requirements. In the context of AM, these steps are necessary because, unlike conventional manufacturing, AM leads to specific unique challenges that arise mainly due to its layer-by-layer fabrication approach [30, 31].

Techniques like polishing, grinding, or blasting are used to improve the surface quality and remove layer lines for surface polishing. Studies have been done by Liu et al. to optimize this [32], which use autoencoders and the concept of optical scattering that helps analyze surface data to predict and detect defects like scratches or rough areas. Also, Mengesha et al. [33] and Snell et al. used clustering algorithms like K-means clustering to analyze surface hardness and internal porosities using 3D and 2D data [33, 34]. The product's quality assessment studies have been done to predict mechanical properties, microstructure analysis, and other physical properties. An example of the prediction of mechanical property lies in the work of Hassanin et al. [35], who used deep learning models to predict mechanical properties like strength and elasticity of lattice structures in Ti-6Al-4V parts, achieving high accuracy [36]. In an interesting experiment, Pazhamannil et al. used artificial neural network (ANN) models to predict tensile strength in FDM-fabricated polylactic acid (PLA) parts [19, 37].

2.3.2.4 Recent developments in material processes and AI advancements

Recent developments in ML within AM highlight a revolutionary change that enhances design optimization, process efficiency, and material properties. Integrating ML models and algorithms revolutionizes how AM processes are interpreted and executed, paving the way for innovative applications and improved outcomes. This section will discuss various aspects of how recent advancements of ML in AM have worked wonders. This section will focus on design optimization techniques, mechanical property prediction, and how the physics-integrated model has helped in this domain.

2.3.2.5 Design optimization

The rapid evolution of ML integration in AM in the design process in the modern world has ultimately been a meaningful change. Implementing advanced ML techniques, such as generative models, has shown enormous potential in designing the architecture of AM models. One notable approach utilizes StyleGAN, a generative adversarial network, to create innovative material designs inspired by natural structures. StyleGAN generates unit cell geometries inspired by natural structures, such as leaf microstructures. Using this technology, highly optimized structures tailored according to various needs can be made, which opens the room for high customization and many innovations [21].

The integration of Bayesian optimization (BO) [21] with variational autoencoders (VAEs) provides an innovative framework for designing composite mechanical metamaterials in AM. This approach is practical for complex multi-material design problems, where traditional trial-and-error methods may be inefficient and time-consuming. BO is a probabilistic optimization method well-suited for expensive-to-evaluate objective functions. VAEs are a generative model that encodes high-dimensional data into a compact latent space representation and then decodes it back to the original space. Now, this integration becomes very handy as the VAE provides a compact and structured latent space where BO can perform optimization more effectively and optimized designs are manufactured.

2.3.2.6 Mechanical properties prediction

Several studies have discussed the application of ML techniques to enhance and predict the mechanical properties of AM products by addressing certain complexities, including process parameters and material characteristics. AM involves multiple parameters and variables like thermal history, process stability, and material properties, affecting the mechanical and microstructural properties of products, and such significant variations of variables are very well connected by the ML models. For PBF, models like regression (OLS, Lasso, Ridge) and ANNs have been used to map process parameters to properties such as hardness and tensile strength. Also, several other ML models (K-nearest neighbor (KNN), DT, and SVMs) have been trained to predict the mechanical properties of LPBF parts using the data collected from various published literature, which contains the mechanical properties corresponding to their laser powder bed fusion (LPBF) process parameters [38]. For DED, techniques like CNNs trained on thermal history data predict properties such as ultimate tensile strength (UTS) with reasonable accuracy. While in material extrusion (ME), ML models, including ANN, SVR, and gradient-boosting models, predict polymer properties (e.g., modulus of elasticity and ductility) and optimize parameters for polymer composites like PLA and ABS [37, 38].

2.3.2.7 Physics-informed models

Recent studies have shown that by combining the ML algorithms and physical laws of manufacturing, the development of physics-informed machine learning (PIML) models has been done by integrating neural networks and physical laws to improve accuracy and generalization. As this study is entirely recent, work has only been done with laser metal deposition (LMD) [39, 40]. Although ML implementation in AM has numerous advantages, traditional data-driven ML models must face specific challenges, such as their "black-box" nature, data dependency, and computationally extensive nature in manufacturing applications. PIML tries to overcome these limitations by incorporating physical laws, such as the conservation of mass, energy, and momentum, into the ML model. A study has been done to understand and analyze complex phenomena like temperature evolution during LMD without relying on extensive labeled datasets, which predict temperature very well on the top boundary with a relative error of 2.1% compared to inside the domain near the bottom with a relative error of 3.7% [39].

2.3.2.8 Future of AI in AM

The future of AI in AM is promising and emerging. It has advancements to enhance efficiency, quality, and innovation across various sectors of AM. This section outlines key areas where AI would impact the field of AM in the future.

2.3.2.9 Generative AI in AM

Generative artificial intelligence (GAI) is a new domain that shows promising capabilities but, in this discussion, particularly in AM. The field is still being researched, and few publications exist. GAI can significantly improve AM processes by generating valuable outputs such as texts, images, and computer codes. This capability of GAI allows for more creative and efficient manufacturing processes. GAI can be used to generate innovative designs and prototypes. This capability of GAI makes the road for more creative solutions, leading to unique and optimized designs that can help improve functionality and performance. GAI can identify existing problems in manufacturing processes by analyzing large amounts of data. It can suggest modifications that help to smoothen operations, reduce waste, and enhance productivity. A recent study showed that it also allows the digital recording and usability of process information, improving overall efficiency [41].

2.3.2.10 Closed-loop AI systems

Closed-loop 3D printing means a system that integrates real-time monitoring and feedback mechanisms to optimize the AM process. Recently, studies have shown that the integration of AI in this area has produced excellent results. AI monitors the AM process continuously and helps find defects and anomalies during the printing operation. Its integration with AI has provided real-time feedback on various printing parameters, including temperature, print speed, and material flow. This feedback loop helps in immediate adjustments, ensuring the printing process remains efficient [42].

2.3.2.11 Transfer learning

Transfer learning is used when the available dataset is insufficient for training the model. It takes or adapts the knowledge from another heavily trained model and then applies it to the required model with a smaller dataset. It assumes common patterns between tasks, enabling knowledge transfer through instance-based, feature-based, model-based, and relational transfer. The AM processes are usually complex and remarkably diverse, so new data is needed for every product. So, active research is being done to implement the concept of transfer learning in this AM domain by reusing knowledge from similar tasks, which reduces the need for extensive data and training. This accelerates the learning and enhances model generalization in AM contexts [43, 44]. Recent studies by Cheng et al. [44] used TL to predict shape deviations of AM products, reducing the need to manufacture new test parts. This field is very emerging, and future research should focus on integrating diverse knowledge types to improve AM modeling.

2.3.2.12 Challenges and limitations

While integrating ML in AM offers immense advantages, certain loopholes and challenges must be considered. These challenges arise from the complexity of AM processes, the diverse types of data and datasets involved, and the need for models that balance accuracy with interpretability. One of the biggest challenges is the collection of large amounts of data, which would be needed for complex model training. AM is a multifaceted, stratified production system shaped by several systems and processes. To improve ML performance, it is necessary to integrate diverse data types (e.g., structured and image data) and use techniques like feature extraction or data fusion to reduce dimensionality. This will enhance model accuracy. The collection of such a broad dataset would be both economically and computationally exhaustive, as it would require significant resources to gather and store large amounts of data from various stages of the AM process. Moreover, obtaining high-quality, diverse datasets for training ML models demands considerable time and investment. Also, regarding the present scenario of AM technology, ML has laid its landmark in only three AM technology classes: PBF, DED, and material extrusion. So, there is more potential for applying ML in other branches of AM.

2.4 SUMMARY

The integration of AI with AM is transforming our approach to manufacturing. By merging the flexibility of AM with the decision-making and predictive capabilities of AI, these technologies create more efficient, precise, and innovative processes. AI facilitates intelligent designs, enhances control during production, and improves the quality of finished products.

Despite persistent challenges, such as the requirement for extensive datasets and significant computational resources, the progress is noteworthy. Emerging methods, including generative models and real-time feedback systems, pave the way for more sustainable and scalable manufacturing practices. With continuous research and innovation, the collaboration between AI and AM has the potential to revolutionize industries, unlocking possibilities that are just beginning to be explored. This partnership marks a significant advancement toward the future of intelligent and automated production.

REFERENCES

1. Patel S, Liu Y, Siddique Z, Ghamarian I. Metal additive manufacturing: principles and applications. *Journal of Manufacturing Processes* 2024;131:1179–201. https://doi.org/10.1016/J.JMAPRO.2024.09.101.
2. Bharat N, Jain R, Bose PSC. A comprehensive overview on additive manufacturing processes: materials, applications, and challenges. *Lecture Notes in*

Mechanical Engineering 2024:95–105. www.springerprofessional.de/en/a-comprehensive-overview-on-additive-manufacturing-processes-mat/26214886

3. Palaniappan S, Joshi SS, Sharma S, Radhakrishnan M, Krishna KVM, Dahotre NB. Additive manufacturing of FeCrAl alloys for nuclear applications – a focused review. *Nuclear Materials and Energy* 2024;40:101702. https://doi.org/10.1016/J.NME.2024.101702
4. Liao H, Zhang T, Li C, Tang C, Li Y, Chen P. Advancements in the research of high- temperature gas-cooled reactor fuel via additive manufacturing techniques. *International Journal of Advanced Nuclear Reactor Design and Technology* 2024;6:14–20. https://doi.org/10.1016/J.JANDT.2024.05.005
5. Alami AH, Ghani Olabi A, Alashkar A, Alasad S, Aljaghoub H, Rezk H, et al. Additive manufacturing in the aerospace and automotive industries: recent trends and role in achieving sustainable development goals. *Ain Shams Engineering Journal* 2023;14:102516. https://doi.org/10.1016/J.ASEJ.2023.102516
6. Karuthapandi S, Murugan R. An experimental investigation of flat wire electrodes and their weld bead quality in the FCAW process. *High Temperature Material Processes* 2017; 21(1), 65–79.
7. Rahman MA, Kumar SR, Selvakumar AS. Significance of additive manufacturing in aerospace and automotive industries. In: *Advances in Additive Manufacturing*. Wiley; 2024, pp. 293–315. https://doi.org/10.1002/9781394238316.CH17
8. SpaceX Reveals 3D-Printed Rocket Engine Parts. n.d. https://designnews.com/3d-printing-software/spacex-reveals-3d-printed-rocket-engine-parts (accessed January 4, 2025).
9. Hoffmann M, Elwany A. In-space additive manufacturing: a review. *Journal of Manufacturing Science and Engineering* 2023;145:020801. https://asmedigitalcollection.asme.org/manufacturingscience/article/145/2/020801/1146197/In-Space-Additive-Manufacturing-A-Review
10. Ye J, El Desouky A, Elwany A. On the applications of additive manufacturing in semiconductor manufacturing equipment. *Journal of Manufacturing Processes* 2024;124:1065–79. https://doi.org/10.1016/J.JMAPRO.2024.05.054
11. Vasishta BV, Preetham TH, Akhil UV, Radhika N. An overview of the evolution of additive manufacturing from 3d to 5d printing in biomedical applications. *Progress in Additive Manufacturing* 2024;1:1–17. https://doi.org/10.1007/S40964-024-00831-6/FIGURES/6
12. Mahapatra S, Victoria SIG, Aich S. Various processing routes for biocompatible materials. In: *Surface Engineering of Biomaterials*. Boca Raton: CRC Press; 2024, pp. 155–74. https://doi.org/10.1201/9781003429920-10
13. Donose R, Maalderink H, Reischle G, Van Der Blij F. Additive manufacturing in ASML lithography machines: benefits and quality assurance. n.d.
14. Karunakaran KP, Suryakumar S, Pushpa V, Akula S. Low cost integration of additive and subtractive processes for hybrid layered manufacturing. *Robotics and Computer-Integrated Manufacturing* 2010;26:490–9. https://doi.org/10.1016/J.RCIM.2010.03.008
15. Zhou L, Miller J, Vezza J, Mayster M, Raffay M, Justice Q, et al. Additive manufacturing: a comprehensive review. *Sensors* 2024;24:2668. https://doi.org/10.3390/S24092668

16. Casalino G, Karamimoghadam M, Contuzzi N. Metal wire additive manufacturing: a comparison between arc laser and laser/arc heat sources. *Inventions* 2023;8:52. https://doi.org/10.3390/INVENTIONS8020052
17. Evans SI, Wang J, Qin J, He Y, Shepherd P, Ding J. A review of WAAM for steel construction – manufacturing, material and geometric properties, design, and future directions. *Structures* 2022;44:1506–22. https://doi.org/10.1016/J.ISTRUC.2022.08.084
18. Osipovich K, Kalashnikov K, Chumaevskii A, Gurianov D, Kalashnikova T, Vorontsov A, et al. Wire-feed electron beam additive manufacturing: a review. *Metals* 2023;13:279. https://doi.org/10.3390/MET13020279
19. Taghizadeh M, Zhu ZH. A comprehensive review on metal laser additive manufacturing in space: modeling and perspectives. *Acta Astronautica* 2024;222:403–21. https://doi.org/10.1016/J.ACTAASTRO.2024.06.027
20. Kumar S, Gopi T, Harikeerthana N, Gupta MK, Gaur V, Krolczyk GM, et al. Machine learning techniques in additive manufacturing: a state of the art review on design, processes and production control. *Journal of Intelligent Manufacturing* 2022;34:21–55. https://doi.org/10.1007/S10845-022-02029-5
21. Yao X, Moon SK, Bi G. A hybrid machine learning approach for additive manufacturing design feature recommendation. *Rapid Prototyping Journal* 2017;23:983–97. https://doi.org/10.1108/RPJ-03-2016-0041
22. Zhou HR, Yang H, Li HQ, Ma YC, Yu S, Shi J, et al. Advancements in machine learning for material design and process optimization in the field of additive manufacturing. *China Foundry* 2024;21:101–15. https://link.springer.com/article/10.1007/s41230-024-3145-3
23. Jiang J, Xiong Y, Zhang Z, Rosen DW. Machine learning integrated design for additive manufacturing. *Journal of Intelligent Manufacturing* 2022;33:1073–86. https://link.springer.com/article/10.1007/s10845-020-01715-6
24. Sigmund O, Maute K. Topology optimization approaches: a comparative review. *Structural and Multidisciplinary Optimization* 2013;48:1031–55. https://link.springer.com/article/10.1007/s00158-013-0978-6
25. Wang C, Tan XP, Tor SB, Lim CS. Machine learning in additive manufacturing: state-of-the-art and perspectives. *Additive Manufacturing* 2020;36:101538. https://doi.org/10.1016/J.ADDMA.2020.101538
26. Goodfellow IJ, Pouget-Abadie J, Mirza M, Xu B, Warde-Farley D, Ozair S, et al. Generative adversarial nets. *Advances in Neural Information Processing Systems* 2014;27:2672–80.
27. Tapia G, Khairallah S, Matthews M, King WE, Elwany A. Gaussian process-based surrogate modeling framework for process planning in laser powder-bed fusion additive manufacturing of 316L stainless steel. *International Journal of Advanced Manufacturing Technology* 2018;94:3591–603. https://link.springer.com/article/10.1007/s00170-017-1045-z
28. Xiong J, Zhang G, Hu J, Wu L. Bead geometry prediction for robotic GMAW-based rapid manufacturing through a neural network and a second-order regression analysis. *Journal of Intelligent Manufacturing* 2014;25:157–63. https://link.springer.com/article/10.1007/s10845-012-0682-1
29. Ye D, Hong GS, Zhang Y, Zhu K, Fuh JYH. Defect detection in selective laser melting technology by acoustic signals with deep belief networks. *International*

Journal of Advanced Manufacturing Technology 2018;96:2791–801. https://link.springer.com/article/10.1007/s00170-018-1728-0

30. Kwon O, Kim HG, Ham MJ, Kim W, Kim GH, Cho JH, et al. A deep neural network for classification of melt-pool images in metal additive manufacturing. *Journal of Intelligent Manufacturing* 2020;31:375–86. https://link.springer.com/article/10.1007/s10845-018-1451-6
31. Chen K, Zhang P, Yan H, Chen G, Sun T, Lu Q, et al. A review of machine learning in additive manufacturing: design and process. *The International Journal of Advanced Manufacturing Technology* 2024;135:3. https://doi.org/10.1007/S00170-024-14543-2
32. Liu M, Senin N, Leach R, P. Lehmann, W. Osten, A. Albertazzi Gonçalves. Intelligent quality monitoring for additive manufactured surfaces by machine learning and light scattering. 2021;2. https://doi.org/10.1117/12.2592554
33. Mengesha BN, Grizzle AC, Demisse W, Klein KL, Elliott A, Tyagi P. Machine learning-enabled quantitative analysis of optically obscure scratches on nickel- plated additively manufactured (AM) samples. *Materials* 2023;16:6301. www.mdpi.com/1996-1944/16/18/6301
34. Summers AJ, Yin H, Fischer RD, Prorok BC, Lou X, He QP. Constructing process maps for pulsed wave laser additive manufacturing with interpretable machine learning. *Journal of Manufacturing Processes* 2023;104:138–49. https://doi.org/10.1016/J.JMAPRO.2023.09.018
35. Hassanin H, Alkendi Y, Elsayed M, Essa K, Zweiri Y. Controlling the properties of additively manufactured cellular structures using machine learning approaches. *Advanced Engineering Materials* 2020;22:1901338. https://doi.org/10.1002/ADEM.201901338
36. Wang H, Li B, Zhang W, Xuan F. Microstructural feature-driven machine learning for predicting mechanical tensile strength of laser powder bed fusion (L-PBF) additively manufactured Ti6Al4V alloy. *Engineering Fracture Mechanics* 2024;295:109788. https://doi.org/10.1016/J.ENGFRACMECH.2023.109788
37. Pazhamannil RV, Govindan P, Sooraj P. Prediction of the tensile strength of polylactic acid fused deposition models using artificial neural network technique. *Materials Today Proceedings* 2021;46:9187–93. https://doi.org/10.1016/J.MATPR.2020.01.199
38. Karuthapandi S, Murugan R, Palani PK. Study and analysis of the macrostructure characteristics in FCAW with the use of a flat wire electrode and by optimizing the process parameter using the Taguchi method and regression analysis. *High Temperature Material Processes* 2016;20(3), 197–224.
39. Deshmankar AP, Sesh Challa J, Singh AR, Prakash S, Senior R. A review of the applications of machine learning for prediction and analysis of mechanical properties and microstructures in additive manufacturing. *Journal of Computing and Information Science in Engineering* 2024;24(12):120801 (14 pages). https://doi.org/10.1115/1.4066575
40. Sharma R, Raissi M, Guo YB. Physics-informed machine learning for smart additive manufacturing. *Machine Learning* 2024;1(1): Article 10761, 10.48550/arXiv.2407.10761
41. Faegh M, Ghungrad S, Oliveira JP, Rao P, Haghighi A. A review on physics-informed machine learning for process–structure–property modeling in

additive manufacturing. *Journal of Manufacturing Processes* 2025;133:524–55. https://doi.org/10.1016/J.JMAPRO.2024.11.066
42. Westphal E, Seitz H, Westphal E, Seitz H. Generative artificial intelligence: analyzing its future applications in additive manufacturing. *Big Data and Cognitive Computing* 2024;8;8:74. www.mdpi.com/2504-2289/8/7/74
43. Sani AR, Zolfagharian A, Kouzani AZ. Artificial intelligence-augmented additive manufacturing: insights on closed-loop 3D printing. *Advanced Intelligent Systems* 2024;6:2400102. https://doi.org/10.1002/AISY.202400102
44. Cheng L, Tsung F, Wang A. A statistical transfer learning perspective for modeling shape deviations in additive manufacturing. *IEEE Robotics and Automation Letters* 2017;2. https://doi.org/10.1109/LRA.2017.2713238

Chapter 3

Advancements in functionally graded materials for additive manufacturing

S. Gejendhiran, Rupshree Ozah, and R. Prithivirajan

3.1 INTRODUCTION

The idea of functionally graded materials (FGMs) was initially developed in Japan in the 1980s while studying thermal barrier materials for use in aircraft. Enhancing the endurance of materials under harsh mechanical and thermal circumstances was the first focus, especially for parts like heat exchangers and turbine blades. Gradients in attributes like wear resistance, electrical conductivity, and biocompatibility were added to the notion over time. FGMs have evolved from theoretical concepts to useful, highly engineered materials suitable for a variety of sectors with the introduction of additive manufacturing (AM) technology. Gradient changes in constitution or structure are a characteristic of graded materials, which allow for customized qualities made for a range of uses. Essentially, FGM fabrication has been transformed by the incorporation of AM technologies, including wire arc additive manufacturing (WAAM), selective laser sintering (SLS), and fused deposition modelling (FDM). This integration enables the development of complicated geometric shapes and accurate control over material gradients. In demanding industries like aerospace, biomedical engineering, etc., these developments are essential for producing components with improved mechanical, thermal, and electrical qualities.

The developments in FGMs for AM are examined in this chapter, with an emphasis on aeronautical, biomedical, and other uses. It also deals with processing and characterization issues, as covered in ref. [1] pointing out ways to get around these obstacles.

3.2 FUNCTIONAL CHARACTERISTICS AND COMPUTATIONAL DESIGN APPROACHES OF FGMS

3.2.1 Functional characteristics

AM techniques have made it easier for manufacturers to create FGMs with multifaceted gradients and numerous properties. FGMs are designed to

 DOI: 10.1201/9781003569633-3

deliver smooth changes in material characteristics by varying composition, structure, or porosity throughout a component. These gradual changes improve the functionality of parts under complex loading or environmental conditions, in which conventional homogeneous materials may fail. The following are some prominent FGMs made using AM techniques, along with their functional characteristics and advantages.

3.2.1.1 Stainless steel/Inconel FGMs

These FGMs, which are made using directed energy deposition (DED) and WAAM, display a gradual evolution from stainless steel (SS316L) to nickel-based Inconel alloys.

Functional characteristics: Blends stainless steel's mechanical strength with Inconel's exceptional corrosion resistance.

Mechanical properties: At the graded interface, it accommodates heat stress while offering exceptional tensile strength and ductility.

Thermal capabilities: Inconel's heat resistance allows it to withstand extremely high temperatures, which makes it appropriate for high-temperature settings like heat exchangers.

Corrosion properties: Provides exceptional resistance to corrosion, which is essential for applications involving chemical processing and maritime environments.

Research: The researchers focus on fabricating FGMs using DED with stainless steel 316L and nickel-based alloys—Inconel 625 and Inconel 718. In the first study, DED was employed to deposit alternating layers of 316L and Inconel 625 powders, analyzing the microstructure and residual stress (Figure 3.1). X-ray diffraction (XRD) revealed tensile residual stresses up to 350 MPa in the Inconel-rich zone, with a well-bonded transition minimizing cracking [2].

The second study used a hybrid approach combining DED and CNC machining to fabricate 316L–Inconel 718 FGMs, improving surface finish and dimensional precision. Surface roughness was reduced from ~12 μm to <2 μm post-machining, and hardness increased gradually from ~210 HV (316L) to ~280 HV (Inconel 718), confirming successful functional grading (Figure 3.2). These studies demonstrate the potential of DED-based techniques for producing high-performance FGMs with tailored mechanical properties and enhanced structural integrity for aerospace and energy applications [3].

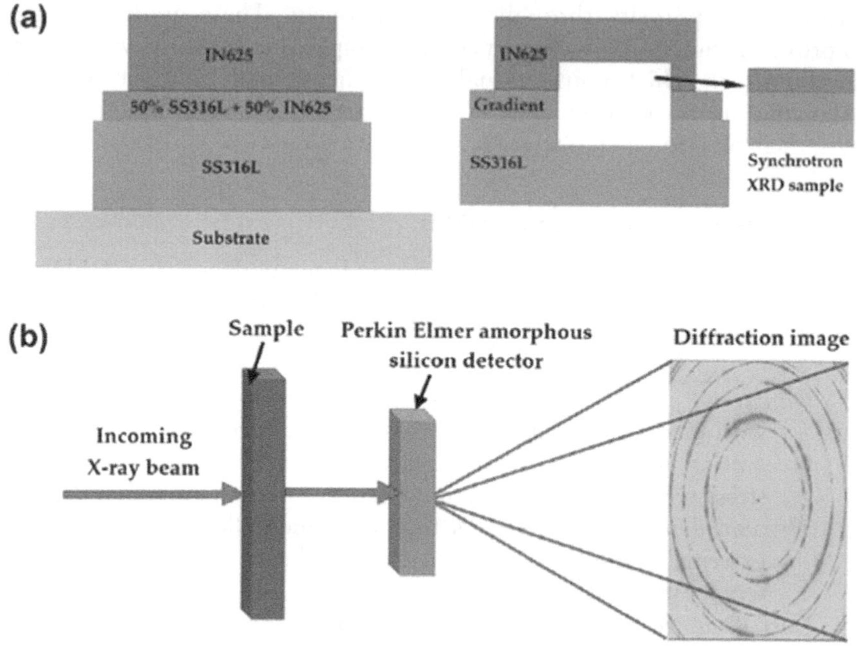

Figure 3.1 Analyzing stress.

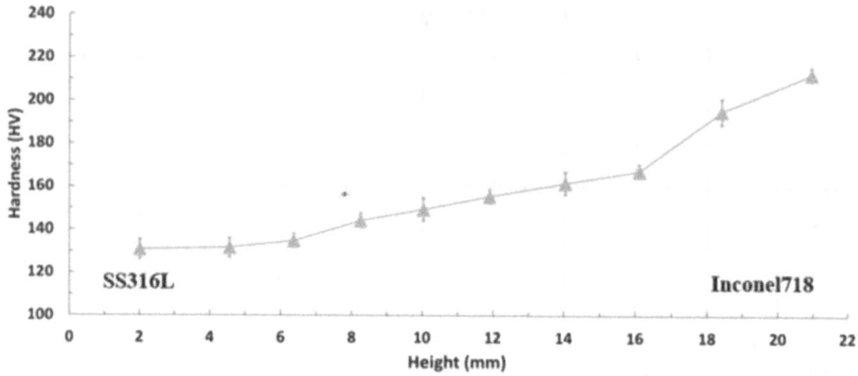

Figure 3.2 Hybrid approach.

3.2.1.2 Titanium/zirconia FGMs

These FGMs are made using laser additive manufacturing (LAM), which combines zirconia's superior wear resistance and bioactivity with titanium's strength and durability.

Functional properties: Provides a gradient in elastic modulus and hardness, making it perfect for load-bearing applications.

Mechanical characteristics: Enhances fatigue resistance and structural integrity by minimizing stress concentrations.

Thermal competencies: Zirconia's heat resistance results in high thermal stability.

Biocompatibility: Zirconia improves tissue compatibility and bone integration, which qualifies these FGMs for orthopedic and dental implants.

Research: The researchers focused the advancements in the fabrication of FGMs using different AM techniques for dissimilar material combinations. The first study employed LAM to produce a gradual transition from Ti6Al4V to zirconia, addressing the inherent challenges of bonding metals with ceramics. The resulting gradient structure exhibited refined grains and stable interfacial phases, with hardness increasing from approximately 350 HV in the metallic region to about 1100 HV in the ceramic zone (Figure 3.3) [4]. In the second study, a hybrid method combining selective laser melting (SLM) and cold spraying (CS) was applied to develop Al–Ti6Al4V FGMs, achieving a strong metallurgical interface and a smooth hardness transition from around 150 HV to 360 HV (Figure 3.4). These investigations confirm the capability of both single and hybrid additive techniques in producing FGMs with reliable structural integrity and customized mechanical properties for demanding engineering applications [5].

3.2.1.3 Polymer-based FGMs

Polymer FGMs, including polylactic acid (PLA) with carbon or ceramic fillers, provide distinctive functional ranges for lightweight and biofunctional components through the incorporation of FDM.

Functional features: Provides customized stiffness and flexibility for industrial and biomedical uses.

Mechanical properties: Enhances load distribution and impact absorption, especially in protective or bio-scaffold applications.

Thermal capabilities: Perfect for areas with moderate heat resistance.

Biocompatibility: PLA-based FGMs promote tissue regeneration and cell proliferation, making them more appropriate for use in bio-implants.

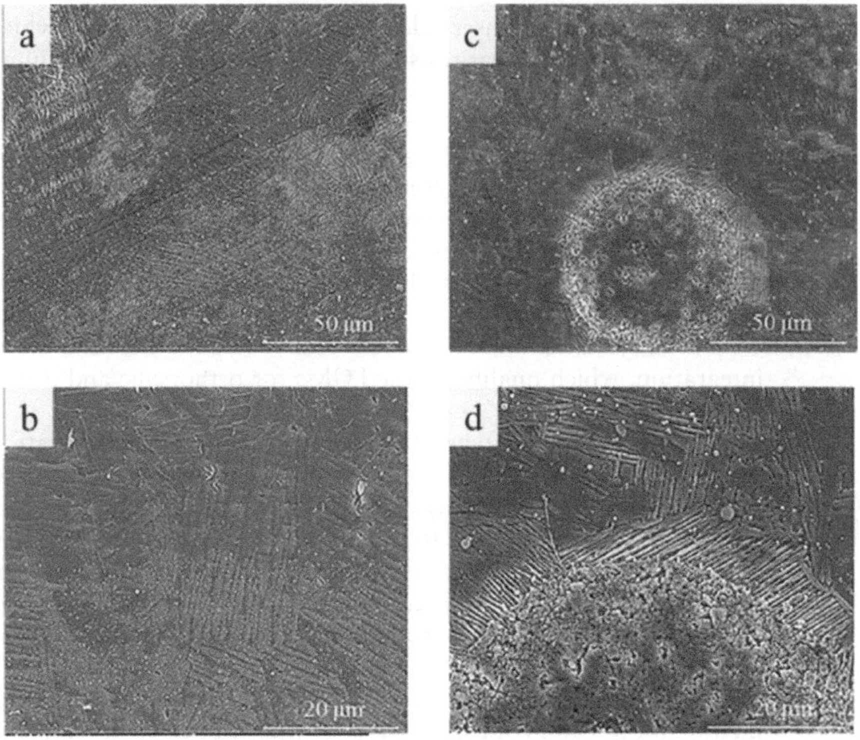

Figure 3.3 Gradient structure.

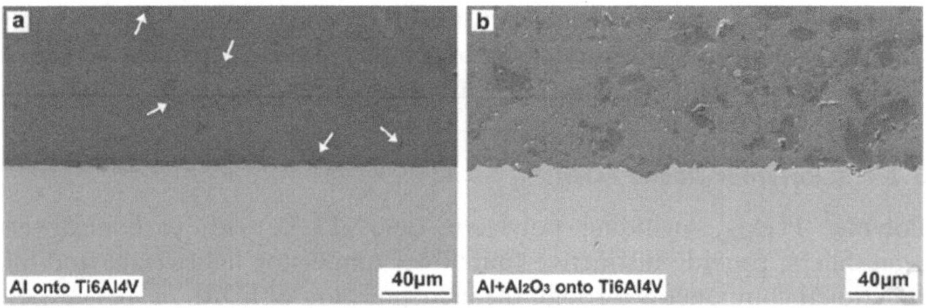

Figure 3.4 Hybrid method of SLM and CS.

Research: The study explored the development of functionally graded thermoplastic composites using FDM combined with coextrusion of continuous fibers. By varying the fiber content across the part during deposition, a tailored gradient in mechanical performance was achieved (Figure 3.5). The material system involved a thermoplastic matrix reinforced with continuous carbon fibers, enabling enhanced strength-to-weight ratios. Mechanical testing showed a significant

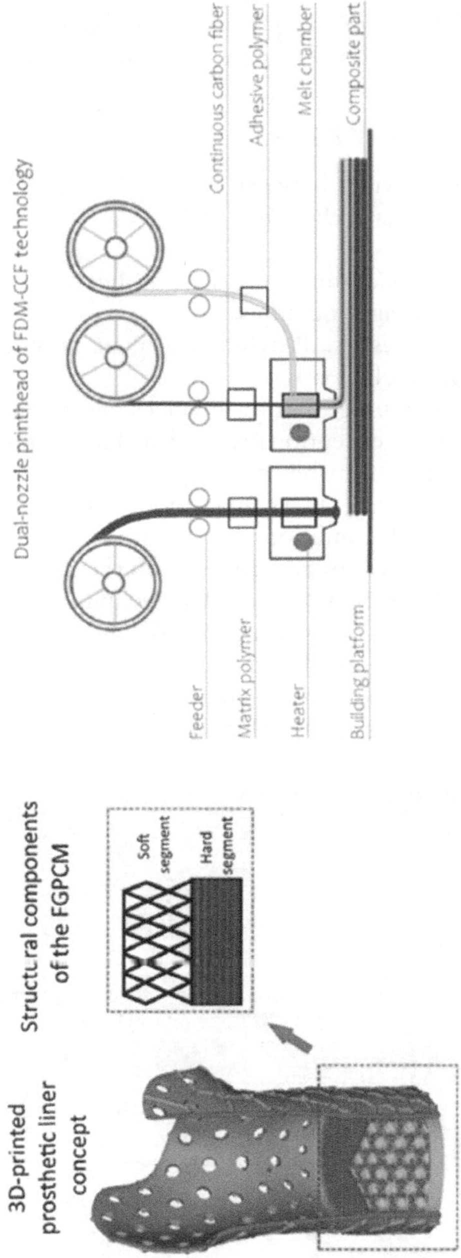

Figure 3.5 Fiber content.

improvement in tensile strength and stiffness in fiber-rich regions, with strength varying from ~40 MPa in low-fiber zones to over 150 MPa in high-fiber regions. This work highlights the potential of coextrusion-based FDM for producing lightweight, high-performance graded composites suitable for aerospace and structural applications [6].

3.2.1.4 Inconel 718/Ti6Al4V FGMs

Fabricating FGMs that switch between Inconel 718 and Ti6Al4V is made easier by laser synchronous preheating (LSP).

Functional features: Combines remarkable heat and corrosion resistance with lightweight qualities.

Mechanical characteristics: Provides excellent fatigue resistance and strength under cyclic loads.

Thermal competencies: Because of Inconel's exceptional thermal resilience, it performs exceptionally well in high-temperature settings, such as aeronautical turbines.

Corrosion characteristics: Offers superior defense against corrosion and oxidation in harsh conditions.

Research: The many researchers focused on the fabrication and evaluation of FGMs using advanced AM techniques. In the first study, a three-phase FGM comprising SS316L, Inconel 625, Inconel 718, and Ti6Al4V was produced via AM. Microstructural analysis showed sound bonding with minimal porosity and distinct grain morphologies across zones (Figure 3.6). Corrosion testing in 5%

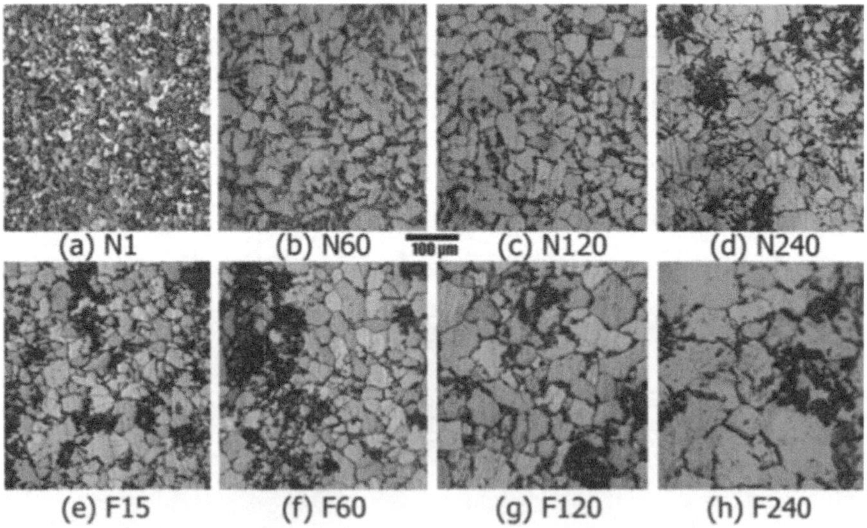

Figure 3.6 Analysis.

NaCl revealed weight losses of ~0.89 mg/cm^2 (SS316L), ~0.32 mg/cm^2 (Inconel 625), ~0.28 mg/cm^2 (Inconel 718), and ~0.45 mg/cm^2 (Ti6Al4V), with pitting mainly in the SS316L region [7]. The second study focused on Ti6Al4V/Inconel 718 FGMs fabricated using directed energy deposition–laser (DED-L). A refined microstructure and good metallurgical bonding were observed, along with intermetallic phase formation at the interface. High-temperature tests up to 800 °C confirmed excellent thermal stability and a consistent hardness of ~400 HV (Figure 3.7). Together, these studies demonstrate the potential of FGMs in enhancing corrosion and high-temperature resistance for advanced engineering applications [8].

3.2.1.5 Metal-ceramic FGMs

FGMs that transition among metals like aluminum as well as ceramics like silicon carbide can be produced using SLS.

Functional features: Designed to meet high mechanical and thermal requirements.

Mechanical properties: Provides exceptional wear resistance and hardness, guaranteeing longevity under extreme strain.

Thermal capabilities: Excellent insulation and resistance to heat shock are provided by this property, which is essential for uses such as thermal barrier coatings.

Properties of corrosion: Prevents oxidation in hot conditions, prolonging component life

Research: The researcher investigated the sinterability of a multi-material system comprising SS316, SiC, and TiN using SLS to fabricate metal–ceramic FGMs. Microstructural analysis revealed partial melting of SS316 and good bonding with ceramic phases in optimized zones. However, thermal mismatch led to porosity and minor cracking in transition regions (Figure 3.8). A relative density above 90% was achieved in well-sintered areas, and hardness increased from ~220 HV (SS316) to ~800 HV in SiC/TiN-rich regions. These results indicate that SLS is a viable method for producing metal–ceramic FGMs with tailored properties [9].

The next research focused on the thermal fracture behavior of metal–ceramic FGMs subjected to steep temperature gradients. Experimental data showed that FGMs with a gradual composition profile experienced up to 40% lower thermal stress compared to sharp-interface composites. Critical temperature differences (ΔT) before fracture increased from ~400 °C (bimaterial interface) to over 650 °C in graded samples. Fractographic

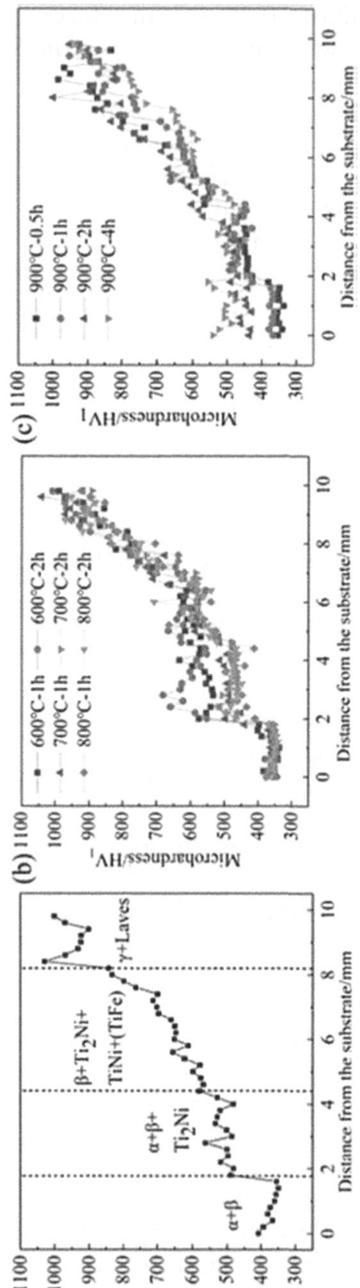

Figure 3.7 High-temperature tests.

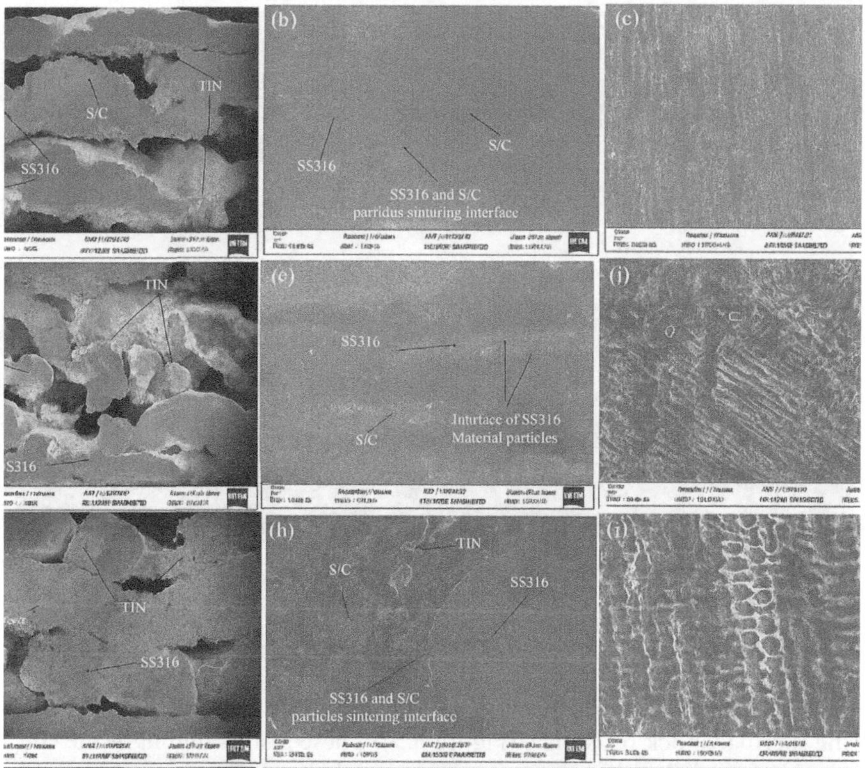

Figure 3.8 Thermal mismatch.

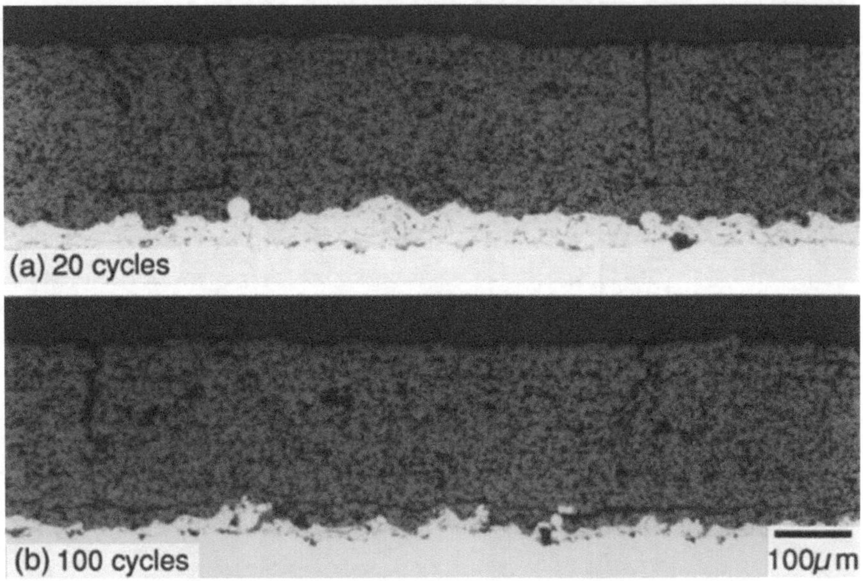

Figure 3.9 Thermal shock resistance.

analysis revealed crack deflection and branching in graded regions, enhancing thermal shock resistance (Figure 3.9). These results highlight the effectiveness of graded architectures in mitigating thermal fracture and improving reliability in high-temperature environments such as aerospace and thermal barrier systems [10].

3.2.2 Computational design approaches

In AM, computational design techniques are essential for the advancement of FGMs. By precisely modeling, optimizing, and simulating material gradients, these techniques guarantee that the finished product will have the intended functional properties. The design of complicated FGMs with customized features can be made easier for researchers and engineers by combining computational techniques with AM techniques. The different computational design approaches are displayed in Figure 3.10.

3.2.2.1 Multi-scale modelling

Multi-scale modeling approaches are crucial for describing the behavior of FGMs at several scales, ranging from macroscopic characteristics to

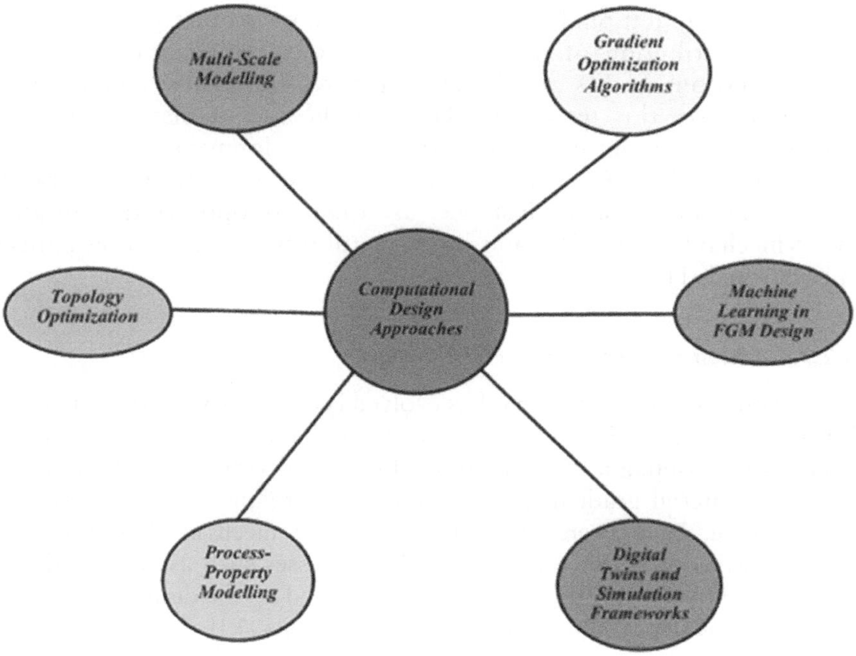

Figure 3.10 Design approaches.

microstructure. Through the simulation of material gradient effects, these models allow for an estimation of thermal, mechanical, and corrosive properties. For example, FGM design has been optimized for temperature resistance and the distribution of stresses with the use of finite element analysis (FEA). In aerospace applications, multi-scale models are very helpful in assessing mechanical and thermal durability in harsh environments [11].

3.2.2.2 *Gradient optimization algorithms*

Advanced computational algorithms, such as genetic algorithms and gradient-based optimization, are employed to design material transitions. These methods optimize the distribution of material phases or compositions to achieve specific functional properties while minimizing defects, such as residual stresses or porosity. For example, gradient optimization has been successfully used to design titanium/zinc-based FGMs with tailored elastic moduli for biomedical implants, ensuring a smooth load transfer and reduced stress shielding.

3.2.2.3 Topology optimization

A potent computational tool that incorporates geometry optimization and the distribution of materials is topology optimization. The creation of porous FGMs and lattice structures for lightweight purposes frequently uses this technique. Through the use of material gradient control, topology optimization enables the development of components with improved thermal management and stiffness-to-weight ratios. Orthopedic implants, for instance, use topology optimization to guarantee mechanical stability and biocompatibility while meeting patient-specific needs [12].

3.2.2.4 Machine learning in FGM design

One revolutionary strategy that has evolved is the incorporation of machine learning (ML) tools into FGM design. Large datasets from simulated and experimental outcomes are analyzed by ML algorithms to find trends, optimize material gradients, and forecast performance. Neural networks, for example, have been used to forecast the mechanical and thermal characteristics of FGMs made with AM processes such as DED and laser powder bed fusion (LPBF). These resources speed up the design process and increase FGM dependability in crucial applications [13].

3.2.2.5 Process-property modeling

Integrating AM process parameters to FGM properties is another goal of computational methods. Researchers can adjust AM procedures to produce ideal material gradients by simulating the relationships between parameters including laser power, scanning speed, and material composition. For instance, LAM process-property modeling has been used to create FGMs with regulated porosity and thermal conductivity properties for thermal barrier coatings in aircraft applications [14].

3.2.2.6 Digital twins and simulation frameworks

Digital twin technology is becoming more popular in the production of FGM because it combines computational models with real-time data from AM operations. By enabling adaptive process control, digital twins guarantee that computational designs are accurately realized. To ensure consistency in final products and minimize trial and error, simulation frameworks also help visualize and validate material gradients prior to manufacture.

The potential of FGMs in AM is largely realized through computational design techniques. Advanced modeling, optimization, and simulation techniques are used to enable engineers to produce devices with previously unheard-of capability and accuracy. From biomedical engineering to the aviation and automotive sectors, these methods provide new possibilities for creative applications.

3.3 ADDITIVE MANUFACTURING TECHNIQUES FOR FUNCTIONALLY GRADED MATERIALS

The manufacturing of FGMs has been transformed by AM techniques, which provide highly customizable material qualities suited to particular applications. This section explores a variety of AM techniques, looking at their underlying theories, their potential for producing FGM, and how to incorporate material gradients. The techniques used for AM of FGMs are displayed in Figure 3.11.

3.3.1 Powder bed fusion AM methods

3.3.1.1 Principle

A fine layer of powder is applied over a building surface using powder bed fusion (PBF) AM techniques. An energy source, such as a laser or electron beam, selectively melts or sinters the powder. The building process takes place layer by layer, with each layer being melted or sintered. To alter characteristics across the layers, material gradients in FGMs are accomplished by changing the powder composition or modifying energy inputs.

3.3.1.2 Production of FGMs by powder bed fusion AM methods

FGMs are produced using PBF techniques like electron beam melting (EBM), SLM, and SLS by carefully adjusting the powder feed or by changing the laser or beam characteristics. This makes it possible to create intricate material gradients that satisfy the demands of various applications, such as high-performance coatings, biomedical implants, and aircraft [15].

3.3.1.3 Techniques

Selective laser melting: SLM uses a laser beam to completely melt metal particles, producing a dense, long-lasting material. Changing the laser's intensity and velocity can change the chemical composition of the material in the same layer, enabling the production of FGMs with specific thermal, mechanical, or electrical characteristics [16].

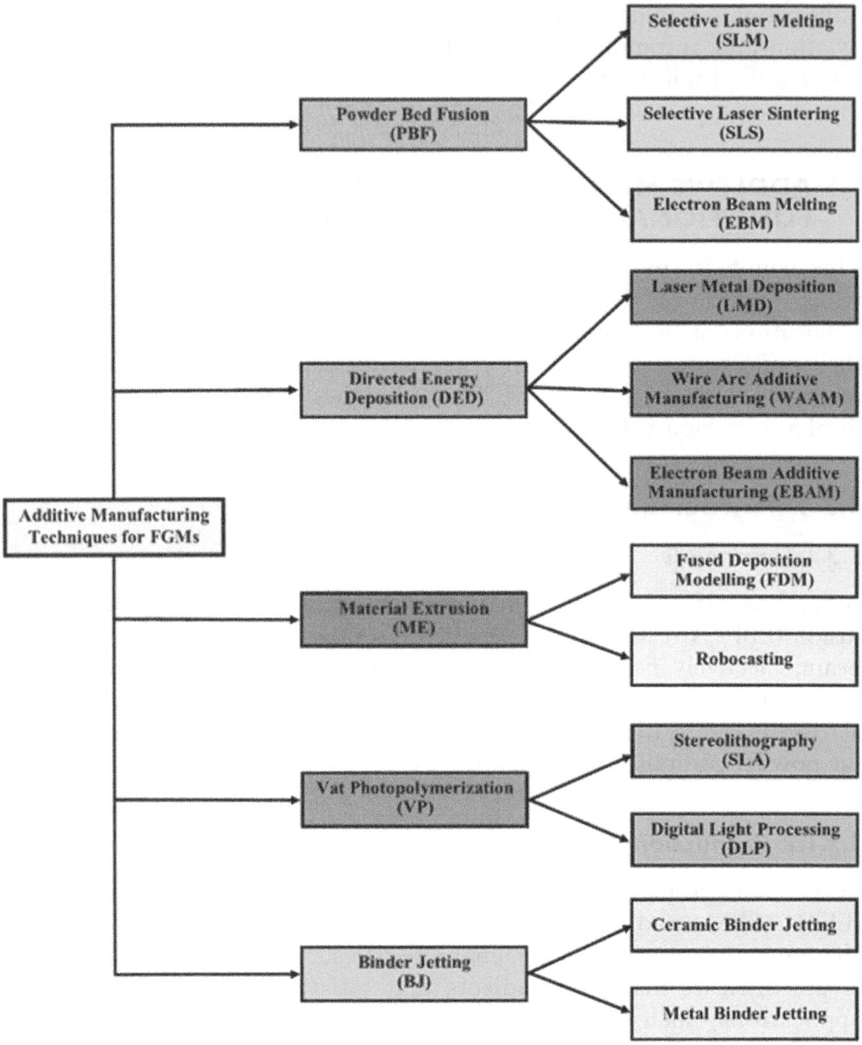

Figure 3.11 Techniques.

Selective laser sintering: This technique involves using a laser to fuse particles of powder collectively at their outermost layer without completely melting them. This technique works best with polymeric and ceramic FGMs, where different sintering techniques or material compositions can produce structures with graded density or porosity that are perfect for energy-absorbing or lightweight applications [17].

Electron beam melting: This technique is especially well-suited for high-temperature alloys like titanium and Inconel because it uses an electron beam to completely melt metal powders in a vacuum environment. The technique creates FGMs by modifying the processing settings or powder constitution to generate materials with improved corrosion and thermal resistance, which are critical for the power generation and aerospace sectors [18].

3.3.2 Directed energy deposition AM methods

3.3.2.1 *Principle*

In DED AM techniques, a material feedstock is exposed to a high-energy source, such as a laser, electron beam, or plasma arc, which melts and deposits the material layer by layer. Real-time changes to the deposition settings or the material flow during the printing process produce material gradients in FGMs.

3.3.2.2 *Production of FGMs by DED AM methods*

The manufacture of FGMs is made easier by DED techniques such as laser metal deposition (LMD), WAAM, and electron beam additive manufacturing (EBAM), which use various feedstock materials or accurately manage material feed rates. Large-scale components that need high-performance materials with specific mechanical properties benefit most from these methods.

3.3.2.3 *Techniques*

Laser metal deposition: LMD melts metal particles with a laser and deposits them on a surface at the same time. FGMs with distinct material qualities can be produced in a single section by altering the powder's concentration or feed rate. This is especially helpful for producing functionally graded coatings or repair applications, including coating turbine blades with wear-resistant layers or producing gradient materials for structural elements [19].

Wire arc additive manufacturing: This process melts metal wire using an electric arc, then deposits the wire layer by layer. Large-scale parts with intricate geometries, such as parts for the aerospace industry and naval applications, are best produced with this technique. FGMs are produced by mixing two or more materials during the deposition process using dual-wire systems or by changing the wire composition, which permits a slow change in material characteristics throughout the part.

Electron beam additive manufacturing: This process uses an electron beam to melt metal wire feedstock in a vacuum. High-performance materials like titanium and super alloys can be used in the process to create FGMs. Complex products with precisely matched mechanical and thermal properties can be produced using EBAM by modifying the wire feed composition and deposit conditions [20].

3.3.3 Material extrusion AM methods

3.3.3.1 Principle

Material extrusion (ME) AM techniques entail layer-by-layer extrusion of a material over a build platform utilizing a heated nozzle. Through material mixing or alternating throughout the extrusion process, this technique can create FGMs, allowing for the production of parts with varying mechanical and thermal qualities throughout.

3.3.3.2 Production of FGMs by material extrusion AM methods

Robocasting and FDM are two examples of ME techniques that enable the creation of objects with graded characteristics. The ability to combine filaments or extrude multiple materials allows for the production of composites for thermal management systems or bio-implants with varying porosities [21].

3.3.3.3 Techniques

Fused deposition modelling: This popular AM method builds components layer by layer by extruding thermoplastic materials using a nozzle that is heated. Following deposition, multiple filament types are blended or alternated to create FGMs. Applications such as graded bio-scaffolds in medical implants or parts with different stiffness can benefit from this technology [11].

Robocasting: This process builds components layer by layer by extruding ceramic slurries. It works especially well for creating FGMs with different material densities or a regulated porosity gradient. Graded mechanical qualities are crucial for performance in high-stress situations in applications such as thermal insulators and structural ceramics [22].

3.3.4 Vat photopolymerization AM methods

3.3.4.1 Principle

Vat photopolymerization is the process of selectively curing a liquid photopolymer resin layer by layer using a UV light source or laser. By changing the resin constitution or the exposure intensity throughout the process of curing, material variations are created, giving exact control upon the mechanical, optical, or thermal properties of the material.

3.3.4.2 Production of FGMs by vat photopolymerization AM methods

FGMs with finely defined features and smooth surface finishes can be produced using methods like stereolithography (SLA) and digital light processing (DLP), which use different resin ingredients or variable light exposure conditions. These techniques work effectively for applications requiring accurate material gradients at the microscale.

3.3.4.3 Techniques

Stereolithography: SLA produces intricate, high-resolution parts by selectively curing photopolymer resin with a laser. Graded dental implants and microfluidic devices are examples of parts with a variety of mechanical qualities that can be created by stacking different resins or altering the curing settings to create FGMs [16].

Digital light processing: DLP offers quicker production speeds than SLA by using digital light projectors to cure a whole layer of resin at once. FGMs can be made with multi-material systems or alternating resin compositions to construct intricate structures with varying qualities that are perfect for practical prototypes or optical lenses [21].

3.3.5 Binder jetting AM methods

3.3.5.1 Principle

A liquid binder is selectively deposited over a bed of powder via binder jetting, which joins the particles to create a solid structure. In the course of the deposition process, FGMs are produced by mixing ingredients in the binder or by layering powders with varying proportions.

3.3.5.2 Production of FGMs by binder jetting AM methods

The creation of FGMs with regulated porosity or different mechanical properties is made possible by the ability to regulate the powder composition and binder dispersion in binder jetting techniques like ceramic and metal. These FGMs are employed in applications that call for lightweight constructions and intricate geometries [19].

3.3.5.3 Techniques

Ceramic binder jetting: This method builds components layer by layer using ceramic powders and binder material. For uses like thermal insulators or high-performance parts subjected to extremely high temperatures, FGMs with different levels of density or thermal conductivity are developed.

Metal binder jetting: This technique uses a selective deposition of binder material to fuse metal powders, resulting in FGMs. For lightweight aircraft constructions or energy-efficient components, the method works well for creating parts with intricate geometries and regulated porosity.

3.4 APPLICATIONS OF ADDITIVE MANUFACTURING OF FGMS AND FGPS

FGMs and functionally graded polymers (FGPs) with specific qualities can be precisely fabricated using AM, providing answers to challenging engineering problems. These materials can satisfy the particular requirements of various industries because of their capacity to regulate material gradients during manufacture. This section examines how both FGMs and FGPs are used in important industries such as electronics, biomedical, energy, automotive, marine, and aerospace, showing that AM improves component performance and usefulness.

3.4.1 Aerospace applications

3.4.1.1 AM of aerospace thermal protection systems (TPS) with FGMs

FGM AM is essential for the aerospace industry's development of thermal protection systems for high-performance airplanes and spacecraft. The composition of these materials varies from thermally insulating components in the core to heat-resistant materials on the surface. Thermal management is optimized through the high-precision fabrication of these FGMs using AM processes including EBM and SLM. The temperature resistance and efficiency of aircraft components during re-entry or high-speed flight are greatly enhanced by this capacity.

3.4.1.2 AM of FGPs in aerospace applications

FGPs are used in aerospace to create components that need to be lightweight and have good performance. FGPs, for instance, can be utilized in specialized sealing components where the inner surface retains its flexibility for sealing capabilities while the outside surface is made to withstand high temperatures. FDM, which enables customized polymer gradients that combine mechanical strength and thermal resistance, is frequently utilized in the manufacturing of FGPs. This improves performance in a variety of aerospace applications [19].

3.4.2 Biomedical applications

3.4.2.1 AM of FGMs in implants and prosthetics

AM of FGMs is transforming implant and prosthetic design in the biomedical industry. Greater compatibility with the neighboring bodily structures is made possible by the ability to modify these materials to comply with the mechanical characteristics of human bone or tissue. FGMs for orthopedic implants are made using processes like SLM and DED, where a gradient structure enables the implant to smoothly change from hard metal at the center to bone-like material at the outermost layer. This speeds up the healing process and lowers the chance of implant failure [23].

3.4.2.2 AM of FGPs in biomedical devices

FGPs, on the other hand, are mainly utilized in smoother biomedical applications like delivery systems for drugs, flexible dressings for wounds, and soft robotics. These kinds of polymers can be produced using techniques like FDM or SLA, which give fine control over the gradient of mechanical characteristics, resulting in making them perfect for devices that need both strength and flexibility. AM enables quick prototyping and customization of biomedical devices, enhancing treatment and recovery effectiveness.

3.4.3 Energy industry applications

3.4.3.1 AM of energy-efficient heat exchangers for FGMs

The production of heat exchangers with graded qualities that maximize thermal control is made possible by the additive fabrication of FGMs in energy systems. AM techniques like SLM or LMD allow for the development of heat exchangers that function well in harsh environments by regulating the gradient from extremely thermally conductive substances to insulating materials. This improves the transmission of energy and lowers the energy dropped as a result of inadequate transfer of heat.

3.4.3.2 AM of FGPs in energy systems

The energy industry also uses FGPs, especially in the manufacturing of flexible and lightweight parts for devices that store electricity. AM of FGPs, for instance, can help batteries and capacitors as the polymer gradient is designed to facilitate both rapid discharge of energy and high-capacity storage. To achieve energy efficiency criteria, these components must be fabricated with accuracy and material property flexibility using AM techniques like FDM and SLA.

3.4.4 Automotive applications

3.4.4.1 AM of FGMs in automobile exhaust systems

The AM of FGMs is helpful to the automobile sector, especially in engine exhaust systems where materials must be able to survive harsh chemical conditions and high temperatures. FGMs with a progressive change in material characteristics from the outermost layer to the core can be created by employing AM techniques like SLM or LMD. When under stress, the core layers offer strength and durability, while the outside layers can be made to withstand heat and corrosion.

3.4.4.2 AM of FGPs in automotive components

FGPs are being increasingly utilized in automotive applications where strong, flexible, and lightweight materials are crucial. Bumpers, dashboard inlays, and seals with different levels of hardness and flexibility are made using AM processes like FDM or SLA. Manufacturers may create parts that satisfy precise performance and cost goals by designing components with gradients of polymer characteristics, which lowers the overall weight of cars and improves fuel efficiency.

3.4.5 Marine applications

3.4.5.1 AM of FGMs in marine coatings resistant to corrosion

FGMs produced using AM are crucial for producing protective coatings in maritime conditions that provide mechanical strength and resistance to corrosion. These coatings' interior layers offer strength and impact resistance, which are essential for use in hulls of ships and offshore platforms, while the outside layers are made to be highly resistant to corrosion. These protective coatings, which can be directly deposited on metal surfaces, are frequently made using methods like SLM and LMD [11].

3.4.6 Electronics and semiconductor applications

3.4.6.1 AM of electronic heat management FGMs

FGMs are essential to the electronics sector, especially in heat-dissipating parts like heat sinks. To dissipate heat, these parts need a material gradient that permits strong thermal conductivity at the surface, whereas the inner layers offer structural support. To maximize heat management in small electronic devices, these components are fabricated using AM processes such as SLM and EBM, which guarantee accuracy in the material gradients.

3.4.6.2 *AM of FGPs in flexible electronics*

FGPs are employed in the production of flexible electronic parts, including display technologies, actuators, and sensors. By using AM techniques like FDM and SLA, these polymer-based components can have their electrical and mechanical characteristics customized. Soft robotics and wearable technology benefit greatly from flexible though robust electronics that can be designed by modifying the gradient structure [19].

3.5 CHALLENGES OF ADDITIVE MANUFACTURING TECHNIQUES IN PROCESSING FGMS

Significant promise exists for the production of FGMs through AM processes. Nonetheless, there are a number of obstacles to this process that must be overcome to completely reap the rewards of FGMs. Material reliability, control of processes, scalability, and post-processing are among these difficulties, and each is essential to producing high-performance and functional parts.

3.5.1 Compatibility of materials

To process FGMs, disparate materials, typically metals, polymers, or ceramics, must be combined into a single structure. It is very difficult to get these materials to work together because:

Thermal mismatch: Internal strains and fractures during cooling or use might result from variations in thermal expansion coefficients.

Chemical interaction: Unwanted material reactions at high processing temperatures might cause interfacial defects or brittle phases.

Material uniformity: It is still difficult to control the transition zones to produce smooth gradients without segregation, especially in procedures that use wire or powder [24].

3.5.2 Regulation of processes

To guarantee a smooth material gradient, AM techniques for FGMs require exact control over the feedstock, energy input, and deposition process. Some of the difficulties are:

Dynamic adjustment: It is challenging to precisely synchronize real-time feedstock or powder composition alterations required by processes like WAAM and LMD.

Thermodynamic gradients: Distortion, residual strains, or irregular microstructures throughout the part might result from excessive thermal input.

Powder handling: Uneven material characteristics can arise from powder segregation or uneven blending in processes such as SLS and SLM [25].

3.5.3 Heterogeneity in microstructure

Microstructural differences are common in FGMs made by AM, particularly in transition zones. These discrepancies result from:

Cooling rates: Defects or irregular grain structures may result from variations in the rates at which different materials solidify.

Formation of porosity: Quick deposition might result in voids or gas trapping, which lowers mechanical performance.

Unintentional phases: Changes in thermal history may result in unintended crystallographic alterations that jeopardize the desired characteristics [26].

3.5.4 Restrictions on geometry and scale

AM is excellent at making complicated shapes; however, there are other challenges when incorporating FGMs into intricate designs:

Scaling issues: Complicated material flow and thermal management make it harder to sustain constant gradients in larger sections.

Multiple gradients: Components that call for multidirectional or overlapping gradients make process planning even more difficult.

3.5.5 Expense and effectiveness

The cost of producing FGMs using AM can be high due to the following factors:

Material costs: FGM-specific feedstocks and powders are substantially more expensive than standard materials.

Longer build times: To guarantee gradient accuracy, methods like SLM and SLA need to be carefully calibrated and have slower deposition speeds.

Post-processing requirements: To improve the characteristics of FGMs, other procedures like heat treatments or surface finishing are frequently needed [27].

3.5.6 Post-processing challenges

Although post-processing adds more complications, it is essential for enhancing the functionality and appearance of FGM components:

Stress relief: Heat treatments need to balance lingering stresses without causing material gradients to deteriorate.
Surface refinement: FGMs with uneven surface hardness find it difficult to undergo processes like polishing or grinding.
Dimensional stability: Post-processing modifications may cause complex or accurate gradient zones to become distorted.

3.5.7 Testing and characterization

It is challenging to assess the quality of FGMs because of their distinct gradient structure:

Gradient assessment: Precise transition analysis requires sophisticated characterization instruments like X-ray tomography and scanning electron microscopy.
Mechanical testing: Specialized approaches are necessary as traditional testing methods could miss subtle changes in material properties.
Defect detection: Nondestructive evaluation techniques still have difficulty identifying internal defects, especially in thick gradient layers [28].

3.5.8 Insufficient standardization

Adoption is made more difficult by the absence of industry-wide standards for FGM manufacturing utilizing AM:

Build guidelines: There are not many frameworks for employing AM techniques to build FGM constructions.
Protocols for processes: Lack of established parameters for feed rates, temperature control, or material transitions.

3.5.9 Safety and environmental issues

Potential risks and environmental effects are associated with AM methods for FGMs:

Material waste: This includes failed projects and unused powders.
Occupational hazards: Inhalation or explosion risks are associated with fine powders used in procedures such as SLM.

3.6 PERSPECTIVES FOR THE FUTURE

3.6.1 Development of advanced feedstocks

New feedstocks designed specifically for FGM production can help with issues like:

Custom powders: Creating bimodal or multimodal powders by combining substances with regulated compositions and particle sizes.
Hybrid feedstock systems: These allow for higher-fidelity gradients through the integration of wires and powders in a single process.

3.6.2 Real-time control and in situ monitoring

Including in situ monitoring devices can greatly increase the accuracy and dependability of FGM production:

Sensors and feedback loops: Deposition variables can be optimized by using real-time temperature, material flow, and composition sensors.
Algorithms for machine learning: AI-powered systems are able to forecast and modify process variables to preserve desirable gradients [13].

3.6.3 Techniques for hybrid manufacturing

There are options to get around restrictions when AM is combined with conventional manufacturing techniques.

Hybrid fabrication: Improving the appearance of surfaces and gradients by combining AM with subtractive techniques like machining.
Functionality integration: Combining smart material systems or integrated electronics with FGM fabrication.

3.6.4 Better tools for computation

The design and modeling of FGMs could be completely transformed by sophisticated computer models:

Algorithms for gradient optimization: Provide exact control over the distribution of materials to reduce flaws.
Microstructure prediction models: Forecast mechanical characteristics and phase formation under various processing environments [29–31].

3.6.5 Extending applications

Diversifying the application of FGMs will spur scalability and innovation:

Energy systems: FGMs are adaptable to fuel cells, heat exchangers, and thermal management systems.
Automotive and marine industries: Premium FGMs can enhance corrosion prevention and wear resistance.

3.6.6 Certification and standardization

Quality and uniformity will be guaranteed by creating defined standards for the manufacture and characterization of FGM:

Global standards: Regulatory agencies, business, and academics work together to establish standards for the production of FGM.
Certification procedures: Guaranteeing the dependability and security of FGM components for vital uses such as biomedical and aeronautical equipment.

3.6.7 Initiatives for environmental sustainability

Long-term viability will be ensured by concentrating on environmentally sustainable AM techniques and materials:

Recyclable feedstocks: Creating materials with many uses.
Energy-efficient methods: Investigating AM techniques with minimal energy consumption or systems that run on renewable energy.

3.7 CONCLUSION

The development of components with precisely customized qualities has been made possible by the revolutionary advancement in material design and production brought about by the incorporation of FGMs into AM techniques. With techniques like FDM, WAAM, and SLS, FGMs enable a gradient distribution of materials that can maximize mechanical, thermal, and electrical properties for particular uses. Significant progress has been made as a result of these skills, particularly in the biomedical and aerospace industries where components need certain qualities to improve functionality, performance, and lifespan. Despite the enormous potential for FGMs in AM, there are still a number of obstacles to overcome. These include high production costs, microstructural irregularities, process control, and material compatibility problems. Furthermore, the broad use of these materials is made more difficult by the lack of uniformity among AM processes for the creation of FGM. It will take ongoing advancements in processing techniques, improved material characterization, and the creation of more effective and economical plans to overcome these obstacles.

Regardless of these challenges, the advancements in FGM processing made possible by AM have the potential to completely transform whole sectors by making it possible to fabricate intricate pieces that were previously impossible to accomplish using conventional manufacturing techniques. The probable future of FGMs in manufacturing is promising as long as research and technology advancements keep improving AM methods' accuracy, effectiveness, and scalability. The integration of FGMs into AM will continue to open up new possibilities and result in more inventive, sustainable, and efficient manufacturing solutions for a variety of applications, provided that continuous efforts are made to improve processing methods and deal with the related difficulties.

REFERENCES

1. R.F. Silva, P.G. Coelho, C.V. Gustavo, C.J. Almeida, F.W.C. Farias, V.R. Duarte, J. Xavier, M.B. Esteves, F.M. Conde, F.G. Cunha, T.G. Santos, Functionally graded materials and structures: Unified approach by optimal design, *Metal Additive Manuf. Image-Based Charact. Mater.* 17 (2024) 4545. https://doi.org/10.3390/ma17184545.
2. X. Zhang, L. Wang, F. Liou, Y. Ren, C. Sun, Microstructure and residual stress in functionally graded 316L stainless steel/Inconel 625 alloys fabricated by direct energy deposition, *JOM* 75 (2023) 5066–5078. https://doi.org/10.1007/s11837-023-06201-x.
3. R. Zhang, K.M. Nagaraja, N. Bian, E. Fisher, S. Ahmadyar, K. Bayazitoglu, H. Lu, W. Li, Experimental studies on fabricating functionally gradient material of stainless steel 316L-Inconel 718 through hybrid manufacturing: directed energy deposition and machining, *Int. J. Adv. Manuf. Technol.* 120 (2022) 7815–7826. https://doi.org/10.1007/s00170-022-09304-y.
4. K. Zhao, G. Zhang, G. Ma, C. Shen, D. Wu, Microstructure and mechanical properties of titanium alloy/zirconia functionally graded materials prepared by laser additive manufacturing, *J. Manuf. Process.* 56 (2020) 616–622. https://doi.org/10.1016/j.jmapro.2020.05.044.
5. S. Yin, X. Yan, C. Chen, R. Jenkins, M. Liu, R. Lupoi, Hybrid additive manufacturing of Al-Ti6Al4V functionally graded materials with selective laser melting and cold spraying, *J. Mater. Process. Technol.* 255 (2018) 650–655. https://doi.org/10.1016/j.jmatprotec.2018.01.015.
6. A. Sotov, A. Abdrakhmanova, A. Zaytsev, A. Popovich, Additive manufacturing of functional gradient thermoplastic composite using fused deposition modelling and coextrusion continuous fibres, *Int. J. Adv. Manuf. Technol.* 137 (2025) 3587–3601. https://doi.org/10.1007/s00170-025-15376-3.
7. K. Sainath, R. Karuppasamy, S. Prabagaran, Investigation of corrosion resistance behavior of additive manufactured SS316L, Inconel 625, Inconel 718 and Ti6Al4V three-phase functionally graded material based on accelerated

weight loss corrosion test using 5% NaCl solution, *Discov. Mech. Eng.* 4 (2025) 11. https://doi.org/10.1007/s44245-025-00095-6.

8. S. Ji, Z. Sun, W. Zhang, X. Chen, G. Xie, H. Chang, Microstructural evolution and high temperature resistance of functionally graded material Ti-6Al-4V/Inconel 718 coated by directed energy deposition-laser, *J. Alloys Compd.* 848 (2020) 156255. https://doi.org/10.1016/j.jallcom.2020.156255.
9. S. Karuthapandi, M. Ramu, P.R. Thyla, Effects of the use of a flat wire electrode in gas metal arc welding and fuzzy logic model for the prediction of weldment shape profile, *J. Mech. Sci. Technol.* 31 (2017) 2477–2486. DOI: 10.1007/s12206-017-0445-1
10. A. Kawasaki, R. Watanabe, Thermal fracture behavior of metal/ceramic functionally graded materials, *Eng. Fract. Mech.* 69 (2002) 1713–1728. https://doi.org/10.1016/S0013-7944(02)00054-1.
11. M. Mohan Krishna Sai, S. Kumar, A. Mandal, M. Anand, Sinterability of SS316, SiC, and TiN multi-material additive manufacturing via selective laser sintering, *Opt. Laser Technol.* 167 (2023) 109686. https://doi.org/10.1016/j.optlastec.2023.109686.
12. M. Teacher, R. Velu, Additive manufacturing of functionally graded materials: A comprehensive review, *Int. J. Precis. Eng. Manuf.* 25 (2024) 165–197. https://doi.org/10.1007/s12541-023-00864-x.
13. D. Mahmoud, M. Elbestawi, Lattice structures and functionally graded materials applications in additive manufacturing of orthopedic implants: A review, *J. Manuf. Mater. Process.* 1 (2017) 13. https://doi.org/10.3390/jmmp 1020013.
14. M. Karimzadeh, D. Basvoju, A. Vakanski, I. Charit, F. Xu, X. Zhang, Machine learning for additive manufacturing of functionally graded materials, *Materials* 17 (2024) 3673. https://doi.org/10.3390/ma17153673.
15. B. Zhang, P. Jaiswal, R. Rai, S. Nelaturi, Additive manufacturing of functionally graded material objects: A review, *J. Comput. Inf. Sci. Eng.* 18 (2018) 041002. https://doi.org/10.1115/1.4039683.
16. K. Sanjeeviprakash, A.R. Kannan, N.S. Shanmugam, Additive manufacturing of metal-based functionally graded materials: overview, recent advancements and challenges, *J. Braz. Soc. Mech. Sci. Eng.* 45 (2023) 241. https://doi.org/10.1007/s40430-023-04174-1.
17. C. Zhang, F. Chen, Z. Huang, M. Jia, G. Chen, Y. Ye, Y. Lin, W. Liu, B. Chen, Q. Shen, L. Zhang, E.J. Lavernia, Additive manufacturing of functionally graded materials: A review, *Mater. Sci. Eng. A* 764 (2019) 138209. https://doi.org/10.1016/j.msea.2019.138209.
18. R. Zhang, F. Jiang, L. Xue, J. Yu, Review of additive manufacturing techniques for large-scale metal functionally graded materials, *Crystals* 12 (2022) 858. https://doi.org/10.3390/cryst12060858.
19. V.A. Popovich, E.V. Borisov, A.A. Popovich, V.Sh. Sufiiarov, D.V. Masaylo, L. Alzina, Functionally graded Inconel 718 processed by additive manufacturing: Crystallographic texture, anisotropy of microstructure and mechanical properties, *Mater. Des.* 114 (2017) 441–449. https://doi.org/10.1016/j.matdes.2016.10.075.

20. A. Reichardt, A.A. Shapiro, R. Otis, R.P. Dillon, J.P. Borgonia, B.W. McEnerney, P. Hosemann, A.M. Beese, Advances in additive manufacturing of metal-based functionally graded materials, *Int. Mater. Rev.* 66 (2021) 1–29. https://doi.org/10.1080/09506608.2019.1709354.
21. T.A. Rodrigues, F.W. Cipriano Farias, K. Zhang, A. Shamsolhodaei, J. Shen, N. Zhou, N. Schell, J. Capek, E. Polatidis, T.G. Santos, J.P. Oliveira, Wire and arc additive manufacturing of 316L stainless steel/Inconel 625 functionally graded material: Development and characterization, *J. Mater. Res. Technol.* 21 (2022) 237–251. https://doi.org/10.1016/j.jmrt.2022.08.169.
22. C. Zhang, Y. Liu, J. Lu, L. Xu, Y. Lin, P. Chen, Q. Sheng, F. Chen, Additive manufacturing and mechanical properties of martensite/austenite functionally graded materials by laser engineered net shaping, *J. Mater. Res. Technol.* 17 (2022) 1570–1581. https://doi.org/10.1016/j.jmrt.2022.01.111.
23. T. Niendorf, S. Leuders, A. Riemer, F. Brenne, T. Tröster, H.A. Richard, D. Schwarze, Functionally graded alloys obtained by additive manufacturing, *Adv. Eng. Mater.* 16 (2014) 857–861. https://doi.org/10.1002/adem.201300579.
24. W. Pompe, H. Worch, M. Epple, W. Friess, M. Gelinsky, P. Greil, U. Hempel, D. Scharnweber, K. Schulte, Functionally graded materials for biomedical applications, *Mater. Sci. Eng. A* 362 (2003) 40–60. https://doi.org/10.1016/S0921-5093(03)00580-X.
25. C. Shen, Z. Pan, D. Cuiuri, J. Roberts, H. Li, Fabrication of Fe-FeAl functionally graded material using the wire-arc additive manufacturing process, *Metall. Mater. Trans. B* 47 (2016) 763–772. https://doi.org/10.1007/s11663-015-0509-5.
26. S.K. Tiwari, S. Pande, S. Agrawal, S.M. Bobade, Selection of selective laser sintering materials for different applications, *Rapid Prototyp. J.* 21 (2015) 630–648. https://doi.org/10.1108/RPJ-03-2013-0027.
27. L. Yan, Y. Chen, F. Liou, Additive manufacturing of functionally graded metallic materials using laser metal deposition, *Addit. Manuf.* 31 (2020) 100901. https://doi.org/10.1016/j.addma.2019.100901.
28. K. Sripriyan, M. Karthigha, A comparative study on the microstructures and mechanical properties of arc welded of HSLA steel, *J. Mech. Sci. Technol.* 38 (2024a) 6791–6798. DOI : 10.1007/s12206-024-1132-7
29. M. Kundu, P. Kadambi, P. Dhatrak, Additive manufacturing of bio-implants using functionally graded materials, in: Malappuram, India, 2022: p. 020062. https://doi.org/10.1063/5.0082039.
30. G.H. Loh, E. Pei, D. Harrison, M.D. Monzón, An overview of functionally graded additive manufacturing, *Addit. Manuf.* 23 (2018) 34–44. https://doi.org/10.1016/j.addma.2018.06.023.
31. I.F. Ituarte, N. Boddeti, V. Hassani, M.L. Dunn, D.W. Rosen, Design and additive manufacture of functionally graded structures based on digital materials, *Addit. Manuf.* 30 (2019) 100839. https://doi.org/10.1016/j.addma.2019.100839.

Chapter 4

An artificial neural network approach of butt joint of Al6061/A588k HSLA steel dissimilar materials

An aerospace application

K. Sripriyan and M. Karthigha

4.1 INTRODUCTION

In recent years, the automotive, aeronautics, and marine industries have widely adopted body-in-white applications, which involve combining different materials to reduce overall vehicle weight and increase fuel efficiency (Sun et al., 2015a). The combinations of these multi-materials can provide high structural stability and good weight reduction, that is, Al alloy/steel (Batistão et al., 2020; Sripriyan et al., 2023). The combination of aluminum alloys and steel in hybrid structures offers significant advantages for weight reduction, fuel economy, and improved performance in dissimilar welding (Li et al., 2018; Liu et al., 2023; Sripriyan et al., 2022). Moreover, the joining of Al alloy and steel remains a challenging task due to differences in their thermophysical properties and the brittle formation of intermetallic compounds (IMCs) at the interface between the two metals (Chen et al., 2019; Belkessa et al., 2018). Many joining techniques are being proposed by many researchers to join Al alloy and steel; for example, adhesives, mechanical joints, fusion, and solid-state joints have successfully been implemented in various levels of applications (Sun et al., 2015a; Karuthapandi et al., 2017).

Among these techniques, the solid-state welding process is often proposed by many researchers for wide applications due to green welding, higher productivity with lower emission of harmful gases (Mirza et al., 2017). The common challenges observed are the formation of intermetallic compound (IMC) at the interface and the distortion of weldments while joining. Compared to the conventional fusion welding process, abovementioned issues are negligible during arc welding processes because the welding torch does not physically contact the substrate material, allowing for more flexible welding and easier automation. Moreover, during the formation of IMC in Al alloy/steel in interface, the most common nonstoichiometric IMC formations are Fe_3Al, FeAl, $FeAl_2$, Fe_2Al_5, and $FeAl_3$ (Sun et al., 2015b). The greater Al content will be problematic in IMCs due to their brittleness, that is, $FeAl_2$, Fe_2Al_5, and $FeAl_3$, and with high iron content, that is, Fe_3Al and

DOI: 10.1201/9781003569633-4

FeAl, might be beneficial due to their specific strength and good wear, oxidation, and corrosion resistance properties (Kulkarni et al., 2019). The IMC type and size of the compounds are mainly influenced by the welding process and its process parameters (Sadeghian et al., 2014). Sun et al. (2015b) investigated into laser welding on butt joints of Al/steel dissimilar materials with ER4043 filler metal. The researchers found the overall thickness of IMCs to be 1.8–6.2 μm at various welding input conditions. Also, the result inferred that Al alloy/steel joint failed at the brazing interface zone and exhibited 120 MPa during the tensile test. Kang et al. (2014) found that the size and distribution of steel particles in the aluminum stir zone (Al-SZ) significantly influence the mechanical properties of AA6181-T4 to HC340LA steel joints, with IMC layers acting as mechanical interlocking, while Karuthapandi et al. (2016) highlighted that defects like voids, influenced by welding and rotational speed, contribute to the deterioration of the mechanical properties of AA1100 to St37 steel joints.

In recent times, the usage of artificial intelligence (AI) has increased rapidly in many industrial applications and it is replacing the demand for physical models (Karuthapandi and Murugan, 2017; https://dl.begellhouse.com/journals/57d172397126f956,050941 0e424f0478,327506f33414f57c.html). Particularly, artificial neural network (ANN) with multiple hidden layers plays an important role in manufacturing industries and is used to accurately create the intelligent system by learning automatically and improving with experience (Abdolrasol et al., 2021). Due to this reason, many researchers have adopted ANNs of their research to predict different characteristics of their innovation by measuring data and training a model by using these datasets (Ali et al., 2019). Abdolrasol et al. (2021) developed a back-propagation neural network algorithm to optimize the resistance spot welding parameters. These techniques are mainly employed for predicting the tensile shear load bearing capacity (TSLBC) value and classifying the quality for resistance spot welding. The result inferred that by using ANN model predicted with 80% training and 20% test set on TSLBC with an accuracy of 87.220% and 92.865%. In the year 2019, Sunphorka et al. (2017) developed a combined ANN and machine vision system to develop an AI-based gas metal arc welding parameter control system. The result represented that the proposed ANN model is suitable to enhance quality, cost-effectiveness, and pre and post preparation for automated applications. Kim et al. (2014) proposed an ANN model based on back propagation (BP) algorithm for modeling of friction stir welding to effectively address the significant effects on mechanical properties of 7075-T6 Al alloy. The result inferred that the predicted values for the hardness and tensile strength are the least mean relative error. Further, the proposed model confirms that the network is adjusted carefully for effective analysis.

As a result of literature studies, the nonlinear modeling will be suitable to predict the characteristics of any complex model. Therefore, this study aimed to address the joining feasibility of dissimilar materials of Al6061 and A588K and examine the bead appearance and morphological characteristics across a range of laser power settings as a primary objective. Further, the study focused on understanding the correlation between process parameters and weld bead characteristics.

4.2 MATERIALS

Due to higher industrial demand to improve the low-cost joining technique for the benefit of the electric vehicle (EV) market by joining aluminum with steel. This paper focuses on the dissimilar joining of A588 grade K high-strength low-alloy (HSLA) to Al6061 aluminum. The above combination of selected base material is widely used in automotive, aerospace industries, and agricultural machinery building (Sripriyan and Karthigha, 2024a, 2024b; Gu et al., 2020), which possesses good ductility, corrosion resistance, high weight-to-strength ratio, excellent weldability, and also exhibits outstanding low-temperature impact toughness (Sampath, 2006).

The selected base materials were prepared as per the dimensional standard of 100 × 50 × 3 mm with a single V-shaped groove with an angle of 60° made as edge preparation and a hot-dip galvanizing process preformat for the HSLA steel (Ali et al., 2019; https://dl.begellhouse.com/journals/57d172397126f956,0509410e424f0478,327506f33414f57c.html). The chemical compositions of selected base materials and filler wire are presented in Table 4.1 as per American Society for Testing and Materials (ASTM) specifications.

4.3 EXPERIMENTAL PROCESSES

The dissimilar jointing of Al and HSLA steel was performed by the laser welding process and by trials conducted to ensure the credibility with selected process parameters from a literature survey. Trials of butt joint were conducted on AWS/ASTM A5.10: Al Si/5 Er4043 of 1.2 mm diameter as filler electrode, and the element composition is presented in Table 4.1 (Mirza et al., 2017). In the pre-processing, 3-mm-thick sheets of Al alloy

Table 4.1 Chemical composition of A588K, A6061, and E308-L in wt.%

Elements	*C (%)*	*Si (%)*	*Mn (%)*	*P (%)*	*S (%)*	*Mg (%)*	*Zn (%)*	*Cr (%)*	*Cr (%)*	*Ni (%)*	*Mo (%)*	*Cu (%)*	*Al (%)*	*Fe (%)*
A588 K	0.102	0.303	0.405	0.081	0.014	–	–	–	054	0.286	0.016	0.3	0.033	Bal
AA6061	–	0.62	0.06	–	–	0.9	0.02	0.17	0.17	–	–	0.28	Bal	0.33
ER4043	–	4.5	0.05	–	–	0.05	0.1	–	–	–	–	0.3	Bal	0.8

Table 4.2 Welding input parameters for MIG/laser welding

Parameters for laser welding	
Process parameter	*Value*
Laser power (kW)	40–120
Welding speed (mm/s)	Manual
Frequency (Hz)	8–12
Pulse distribution (ms)	3–5
Angle of incidence	Normal
Focal plane	Top surface of the plate
Shielding gas	Argon
Gas flow rate (lpm)	20

and HSLA steel were cut to 100 mm × 50 mm × 3 mm size as a standard size by wire-cut process and the surface was cleaned and polished with different sand papers and cleaned with acetone etch to confirm perfect contact between the plates (Scotti and Rodrigues, 2009). Based on the literature survey, laser power, frequency, and pulse distribution are selected as active parameters for laser welding, with the remaining parameters considered as a passive parameters (shown in Table 4.2).

The standard post-process procedure was performed after the welding and the best samples were selected from each trial. For the pilot trials, the laser power source has been considered at three levels: 5–50 kW (L1), 51–100 kW (L2), and 101–150 kW (L3). In addition to that, a constant arc length of 3 mm was maintained during welding and 20 lpm shielding argon gas used to protect the liquid filler metal from oxidation during its cooling so as to improve the melding ability of the weld seam. Overall, 27 trials (three runs with uniform distribution laser power in each group) were performed.

The outcome of laser welding process parameters on weld bead characteristics, that is, top surface appearance and cross-section of the joints obtained, is shown in Figure 4.1. All the weld bead top surfaces are visually observed as good; no defect is found in any of the welded samples, viz., blowholes, uneven penetrations, excessive deposition, poor forming, oxidation blackening, burning-through, collapse, crack, and surface porosity. The results indicated that a good weld bead appearance with deep penetration was obtained in the laser welding process by promoting the maximum weldability and spreadability of the weld metal at a relatively high temperature. However, a few laser-welded samples were observed to have poor weldability and spreadability of the molten metal at lower laser power due to insufficient laser heat input.

In addition, the study revealed that the performance of the weld joint improved by increasing laser power. Conversely, excessive heat input will cause a narrow bead, uneven melting of the Al base metal, and improper

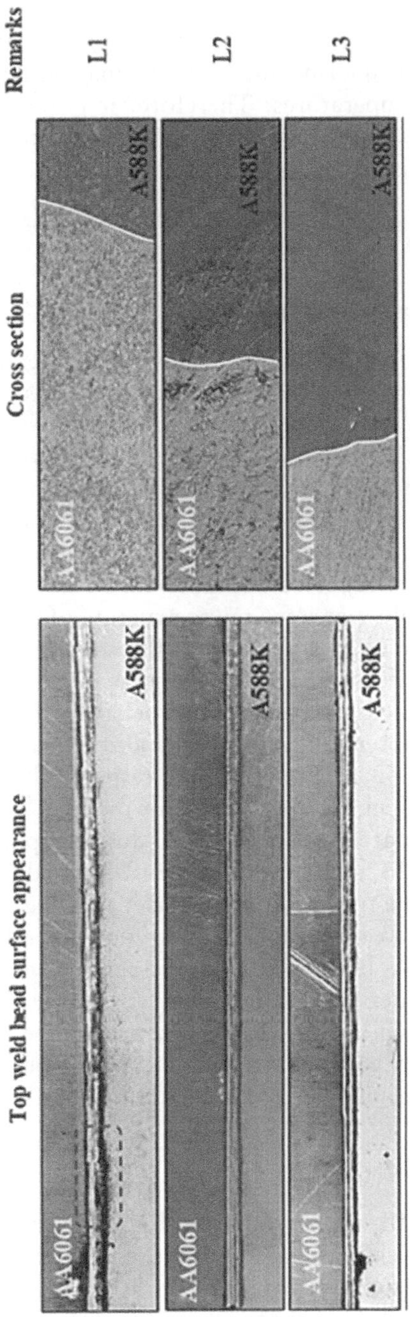

Figure 4.1 Outcome.

spreading of molten filler metal. The common observation for the process is that a perfect weld bead appearance with maximum penetration is obtained at relatively higher temperatures of the molten metal, and the joint strength is relatively based on intermetallic zone formation between molten metal and steel at high temperatures. Therefore, it is necessary to understand the relationship between the welding input parameters and the weld bead characteristics, that is, the metallurgical joint behaviors.

4.4 RESULT AND DISCUSSION

4.4.1 Morphological observation

To exhibit an interfacial formation of dissimilar metals of Al and A588K HSLA steel, the morphological study was conducted. The weld bead top surface and cross-section morphology obtained using laser welding are shown in Figure 4.1. The laser welding is observed as even and small surface and both the welded samples are free from defects, as shown in Figure 4.1. There was an improved cross-sectional surface in laser welding, with the depth of penetration of both samples being the same. The common interfacial observations are reported in Figure 4.2.

In Figure 4.2, L2 illustrates the microstructure of the fusion zone (FZ) on the laser-welded joint of Al and A588K HSLA steel under the heat input of 0.61 kJ/mm. The uniform fine grain is observed with the influence of heat input and amalgamated dissimilar metal behavior. In Figure 4.2, L8, L17, and L20 were also observed as a similar metrological formation under the heat input of 0.38, 0.52, and 0.84 kJ/mm, respectively. However, fine micro cracks, porous and inclusion are noted in the trial cases. Martensite formation at dashed square areas is observed due to rapid solidification in the trial cases L5 and L23 at the heat input of 0.49 and 0.84 kJ/mm, respectively. However, all the trial cases of the laser welding offer enhanced metal pool quality and are free from formation of polygonal ferrite structure (Wei and Nelson, 2011). As both welding processes offer maximum throughput, it is necessary to understand the influence of process parameters on weld characteristics. To achieve this objective, a machine learning (ML) algorithm was developed to evaluate and accurately predict the relationship between input parameters and welding characteristics.

4.4.2 ANN model

In this study, an ML approach is used to qualify the most significant input variables on welding characteristics. In general, ML algorithms are mainly classified as supervised learning (task-driven approach), unsupervised learning (data-driven approach), semi-supervised learning (labeled + unlabeled), and reinforcement learning (environment-driven approach)

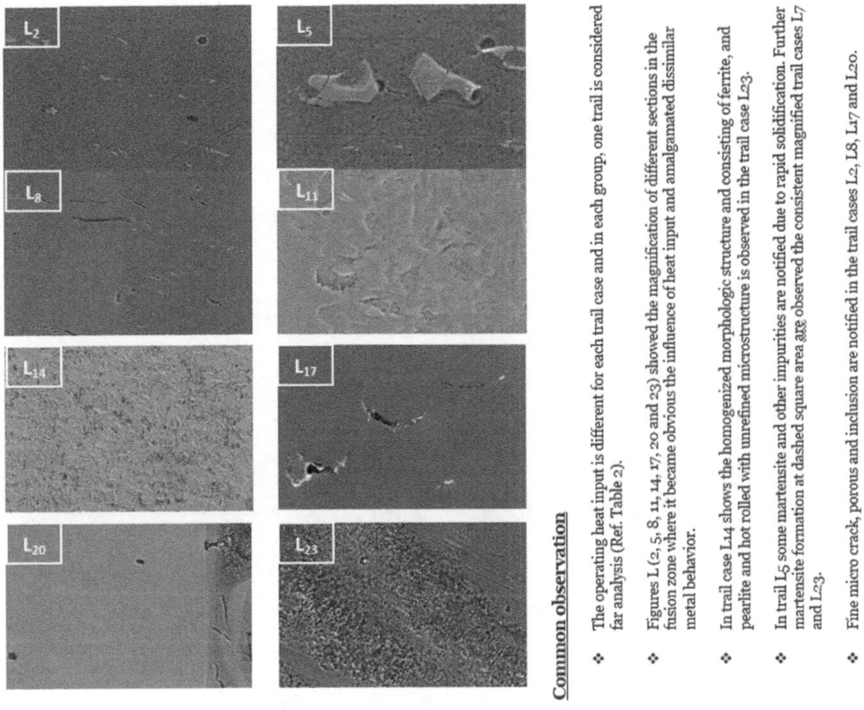

Figure 4.2 Results.

(Soler et al., 2022). In this investigation, task-driven approach is used to label welding datasets to train algorithms to predict input parameters on welding characteristics (Husaini et al., 2020). Based on an extensive literature survey, ML and deep learning techniques, that is, ANN algorithm, are to provide accurate predictions and insights because they learn the characteristics from input–output data and continually re-train with fresh data to adapt to new information (Soler et al., 2022). To understand the performance of the proposed ANN model, the welding (tensile and compressive) data points were split into training data (80%) and the remaining dataset was used to evaluate and predict the output characteristic of the welding. The three layers and the factors included in the final model are presented in Figure 4.3.

Initially the experiments were designed and performed based on full factorial orthogonal design. Overall, 27 trial runs were conducted. In this proposed ANN model, three hidden layers are proposed to understand the influence of welding input parameters and welding characteristics, to ensure the following decision function is expressed as a nonlinear mapping (see Equation 4.1).

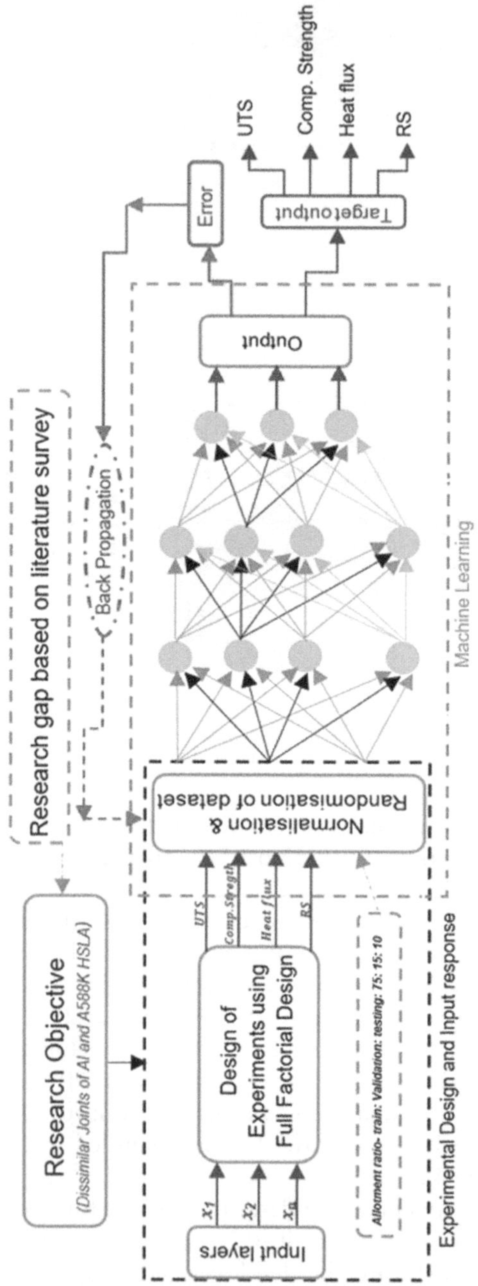

Figure 4.3 Final model.

$$\left(Y^1\right) = f^1\left(w^1x^n + b^1\right)$$ Initial function

$$\left(Y^2\right) = f^2\left(w^2y^1 + b^2\right)$$ Machine learning (hidden layer) functions (4.1)

....

....

$$\left(Y^n\right) = f^n\left(w^ny^{n-1} + b^n\right)$$ Final output function for each characteristic

where Y^n is the welding characteristics, x^n is the welding input parameters, f^n is a nonlinear mapping obtained from each neuron in the second and third layers, w^n denotes weight, and b^n is bias function of the welding input and output.

Then

$$x^n = \begin{bmatrix} x_1 \\ \cdot \\ \cdot \\ x^n \end{bmatrix} \quad b^n = \begin{bmatrix} x_1 \\ \cdot \\ \cdot \\ x^n \end{bmatrix} \quad \text{and} \quad w^n = \begin{bmatrix} w_1 & \cdot & \cdot & w_{1n} \\ \cdot & \cdot & \cdot & \cdot \\ \cdot & \cdot & \cdot & \cdot \\ w_{nr} & \cdot & \cdot & w_{nr} \end{bmatrix}$$

The final output was calculated using binary sigmoidal activation function in the ANN model (logistic function), expressed in Equation 4.2 (Abdolrasol et al., 2021). In general, two different types of training procedures are used to obtain the prediction information and error, which are forward and backward propagation, respectively.

$$f(y) = \frac{1}{1 + e^{-y}} \tag{4.2}$$

Figures 4.4–4.6 describe the training analysis of the laser welding process of an ANN graph; the response indicated the mean squared error (MSE) value over 15, 14 epochs for the training, validation, and test sets. Each model attained its best validation performance, that is, MSE = 1.8879e-09 at epoch 15 for laser welding. The training process successfully reduced MSE for all datasets, stabilizing at a point where additional training improved welding characteristics. The convergence of validation and test MSEs indicates good generalization and minimal overfitting. Overall, the model demonstrates robust performance on both training and unseen data, achieving excellent results based on the MSE metric.

The three plots related to the ANN training process given in Figure 4.6 show the gradient, parameter Mu (μ), and validation checks over 15 and 11 epochs for laser welding. The gradient algorithm helps minimize error

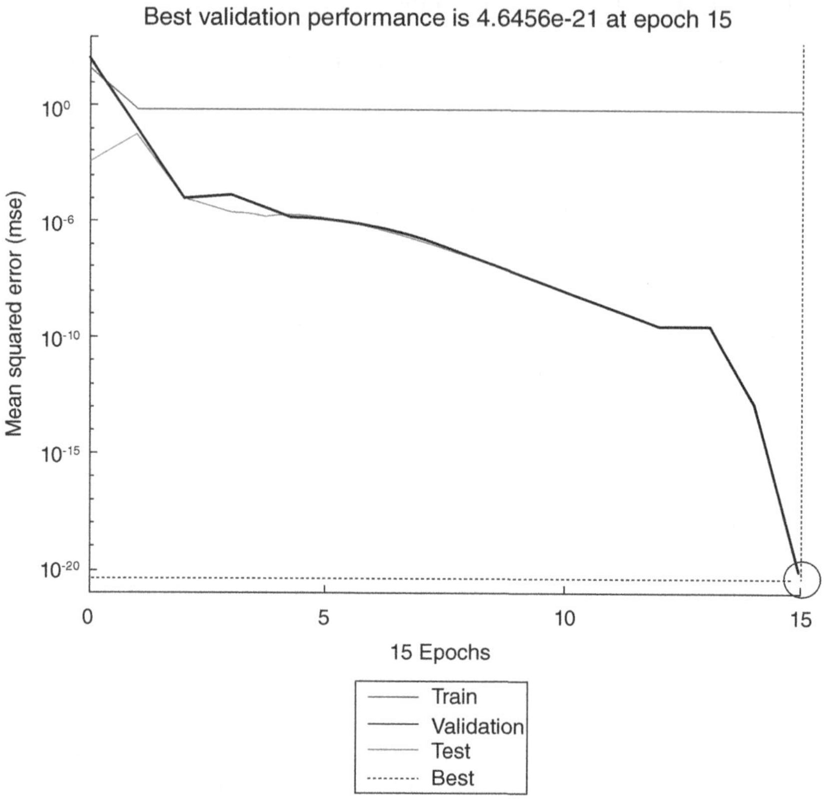

Figure 4.4 Training analysis (I).

by adjusting weights. In the proposed models, the Levenberg–Marquardt backpropagation yielded gradient values of 1.8879e-09 and 0.00012704, with μ parameters of 1e-09 for laser welding. Regression plots (Figure 4.5) indicate excellent performance for both processes, with R-values near 1, showing almost perfect correlation. The fit lines closely align with Output = Target, indicating minimal bias and strong generalization. Overall, the gradient plot demonstrates effective convergence, with the gradient stabilizing at a low value by the end of training.

4.5 CONCLUSIONS

The joining of two dissimilar metals of Al6061 and A588 HSLA steel, was performed using a laser welding process in a butt joint configuration. The key observations from the weld bead surface and metallography are as follows:

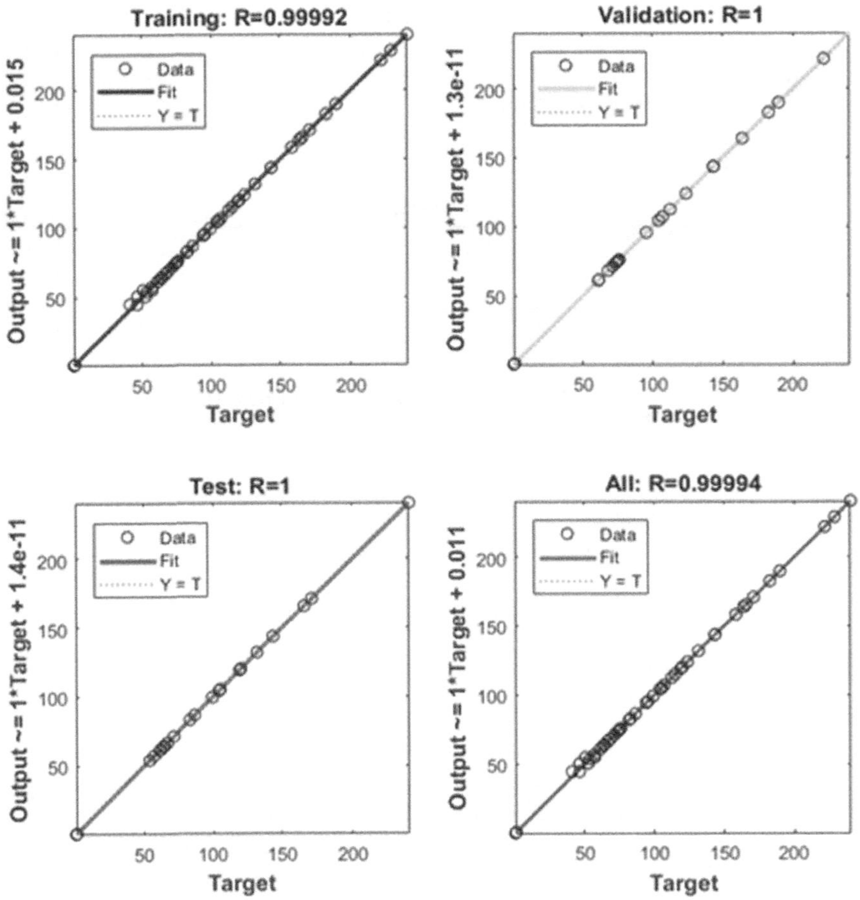

Figure 4.5 Training analysis (II).

- All weld samples exhibit good surface finish and excellent top surface appearance with perfect dispersion at the Al/steel interface layer. The mechanical tests such as tensile and compressive were observed with notable improvements.
- Weld bead top surface and cross-section morphology were studied and the result indicated an even and small surface welding, free from defects. However, uniform fine grain was observed with the influence of heat input (i.e., 0.38, 0.52, and 0.84 kJ/mm) and amalgamated dissimilar metal behavior.

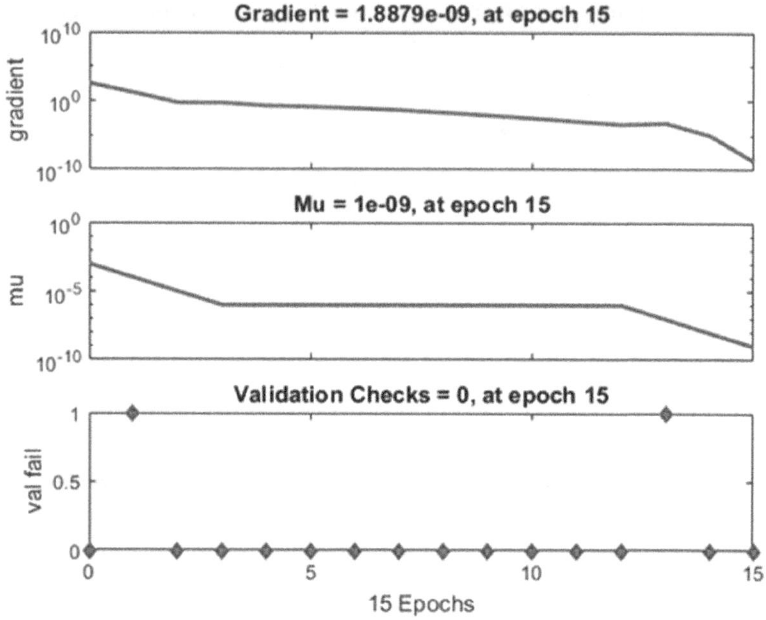

Figure 4.6 Training analysis (III).

- Martensite formation at dashed square areas is observed due to rapid solidification under the heat input of 0.49 and 0.84 kJ/mm, and free from the formation of polygonal ferrite structure
- The common observation was welding bead displayed Al–Zn eutectoid phases and gray α-Al formations between the Al–Zn dendrites. Due to the presence of stabilizing elements in the weld metal, such as nickel and manganese, the steel did not fully transform to Al during cooling in either process.
- ANN technique was used to qualifying the most significant input variables on welding characteristics and the response indicated the MSE value over 15, 14 epochs for the training, validation, and test sets. Each model attained its best validation performance, that is, MSE = 1.8879e-09 at epoch 15. The regression performance of laser process with *R*-values near 1 shows almost perfect correlation. The fit lines closely align with Output = Target, thereby indicating minimal bias and strong generalization. Overall, the gradient plot demonstrates effective convergence, with the gradient stabilizing at a low value by the end of training.

These results demonstrate that both laser welds, under optimized conditions, can achieve high-strength, high-quality Al6061-A588 HSLA steel joints that are suitable for applications requiring durability and impact resistance.

REFERENCES

Abdolrasol, M.G.M. et al. (2021) *Artificial Neural Networks Based Optimization Techniques: A Review*, Switzerland: MDPI Electronics.

Ali, M. et al. (2019) 'The effect of double austenitization and quenching on the microstructure and mechanical properties of CrNiMoWMnV ultrahigh-strength steels after low-temperature tempering', *Materials Science and Engineering A*, 763(April), p. 138169.

Batistão, B.F. et al. (2020) 'Characterization of dissimilar friction stir welded lap joints of AA5083 and GL D36 steel', *Journal of Materials Research and Technology*, 9(6), pp. 15132–15142.

Belkessa, B. et al. (2018) 'Dissimilar welding between 2205 duplex stainless steel and API X52 high strength low alloy steel', *Diffusion Foundations*, 18, pp. 7–13.

Chen, L. et al. (2019) 'Numerical and experimental investigation on microstructure and residual stress of multi-pass hybrid laser-arc welded 316L steel', *Materials and Design*, 168. www.sciencedirect.com/science/article/pii/S0264127519300905

Gu, J. et al. (2020) 'Microstructure, mechanical properties, and residual stress distribution of AISI 316L stainless steel part fabricated by laser metal deposition', *Scanning*. https://onlinelibrary.wiley.com/doi/10.1155/2020/4831798

Husaini, N.A. et al. (2020) 'A modified weight optimization for artificial higher order neural networks in physical time series', *International Journal of Advanced Computer Science and Applications*, 11(3), pp. 299–308.

Kang, S.H., Lee, Y.S. and Lee, H.W. (2014) 'Evaluation of flow stress and damage index at large plastic strain by simulating tensile test of Al6061 plates with various grain sizes', *International Journal of Mechanical Sciences*, 80, pp. 54–61.

Karuthapandi, S. and Murugan, R. (2017) 'An experimental investigation of flat wire electrodes and their weld bead quality in the FCAW process', *High Temperature Material Processes*, 21(1), pp. 65–79.

Karuthapandi, S., Murugan, R. and Palani, P.K. (2016) 'Study and analysis of the macrostructure characteristics in FCAW with the use of a flat wire electrode and by optimizing the process parameter using the Taguchi method and regression analysis', *High Temperature Material Processes*, 20(3), pp. 197–224.

Karuthapandi, S., Ramu, M. and Thyla, P.R. (2017) 'Effects of the use of a flat wire electrode in gas metal arc welding and fuzzy logic model for the prediction of weldment shape profile', *Journal of Mechanical Science and Technology*, 31(5), pp. 65–79.

Kim, I.S. et al. (2014) 'Prediction of GMA welding characteristic parameter by artificial neural network system', *Advanced Materials Research*, 1061–1062, pp. 481–491. https://eprints.qut.edu.au/82718/

Kulkarni, A., Dwivedi, D.K. and Vasudevan, M. (2019) 'Dissimilar metal welding of P91 steel-AISI 316L SS with Incoloy 800 and Inconel 600 interlayers by using activated TIG welding process and its effect on the microstructure and mechanical properties', *Journal of Materials Processing Technology*, 274(June), 116280.

Li, L. et al. (2018) 'Influence of laser power on interfacial microstructure and mechanical properties of laser welded-brazed Al/steel dissimilar butted joint', *Journal of Manufacturing Processes*, 32, pp. 160–174.

Liu, J. et al. (2023) 'Microstructure and mechanical properties of aluminum-steel dissimilar metal welded using arc and friction stir hybrid welding', *Materials and Design*, 225.

Mirza, F.A. et al. (2017) 'Microstructure, tensile and fatigue properties of ultrasonic spot welded aluminum to galvanized high-strength-low-alloy and low-carbon steel sheets', *Materials Science and Engineering: A*, 690, pp. 323–336.

Sadeghian, M., Shamanian, M. and Shafyei, A. (2014) 'Effect of heat input on microstructure and mechanical properties of dissimilar joints between super duplex stainless steel and high strength low alloy steel', *Materials and Design*, 60, pp. 678–684.

Sampath, K. (2006) 'An understanding of HSLA-65 plate steels', *Journal of Materials Engineering and Performance*, 15(1), pp. 32–40.

Scotti, A. and Rodrigues, C.E.A.L. (2009) 'Determination of the momentum of droplets impinging on the pool during aluminium GMAW', *Soldagem & Inspeção*, 14(4), pp. 336–343.

Soler, D. et al. (2022) 'Prediction of surface roughness of SLM built parts after finishing processes using an artificial neural network', *Journal of Manufacturing and Materials Processing*, 6(4), p. 82. www.mdpi.com/2504-4494/6/4/82

Sripriyan, K. et al. (2022) 'Characteristic of weld bead using flat wire electrode in GMAW inline during the process: an experimental and numerical analysis', *International Journal of Pressure Vessels and Piping*, 196(January).

Sripriyan, K. and Karthigha, M. (2024a) 'A comparative study on the microstructures and mechanical properties of arc welded of HSLA steel', *Journal of Mechanical Science and Technology*, 38(12), pp. 6791–6798.

Sripriyan, K. and Karthigha, M. (2024b) 'Influence of granite micro particles on the mechanical, damping and antimicrobial properties of silk–sisal hybrid composites', *Scientific Reports*, 14(1), pp. 1–14.

Sripriyan, K., Ramu, M. and Thyla, P.R. (2023) 'Characterization of dissimilar welding of AISI 409 ferritic to AISI 316l austenitic stainless steel using AISI 308 filler wire under different temperature conditions', in *Advances in Additive Manufacturing and Metal Joining: Proceedings of AIMTDR 2021*, pp. 309–321.

Sun, J. et al. (2015a) 'Investigation of laser welding on butt joints of Al/steel dissimilar materials', *Materials and Design*, 83, pp. 120–128.

Sun, J. et al. (2015b) 'Investigation of laser welding on butt joints of Al/steel dissimilar materials', *Materials and Design*, 83, pp. 120–128.

Sunphorka, S., Chalermsinsuwan, B. and Piumsomboon, P. (2017) 'Artificial neural network model for the prediction of kinetic parameters of biomass pyrolysis from its constituents', *Fuel*, 193, pp. 142–158.

Wei, L.Y. and Nelson, T.W. (2011) 'Correlation of microstructures and process variables in FSW HSLA-65 steel', *Welding Journal*, 90(5).

Chapter 5

Smart design and material optimization strategies for additive manufacturing

S. Manikandan, N. Saravanan, Y. Robinson, and S. Suresh

5.1 INTRODUCTION

Additive manufacturing (AM) is revolutionizing industries by enabling the creation of complex, customized products with reduced material waste. To maximize the potential of this technology, smart design and material optimization strategies are essential. These strategies focus on improving product performance, reducing production costs, and minimizing environmental impact. The ability of organisms to adjust to environmental changes offers valuable insights into the development and creation of innovative smart systems. As requirements increase, the ability of smart materials to change their shapes has become a broader aim beyond their original capabilities. In contrast to conventional manufacturing methods, AM skillfully combines precise three-dimensional (3D) structures and the intricate response mechanisms of biological organisms with smart materials. These smart structures naturally have been designed and built in various organisms, highlighting the relationship between the structural characteristics and patterns of deformation. The review then discusses how smart bionic structures developed using AM techniques respond to different stimulation, Figure 5.1 illustrates application of AM in nano composites. Additionally, the potential uses of smart bionic structures in biomedicine, intelligent robotics, origami construction, and aerospace are discussed. Finally, the challenges and future prospects for smart bionic structures are examined with the goal of offering innovative solutions for creating the next generation of smart systems through interdisciplinary research [1]. Making a prosthetic leg lighter can improve the mobility of a person and reduce the energy required to climb stairs, enhancing their quality of life and independence. Traditional manufacturing methods, such as injection molding and computerized numerical control (CNC) machining, have limitations in terms of design flexibility.

The optimization process focused on the knee, upper leg and lower leg, and hydraulic and oil support system. The workflow procedure for AM was discussed. Overall, the study successfully reduced the weight of

DOI: 10.1201/9781003569633-5

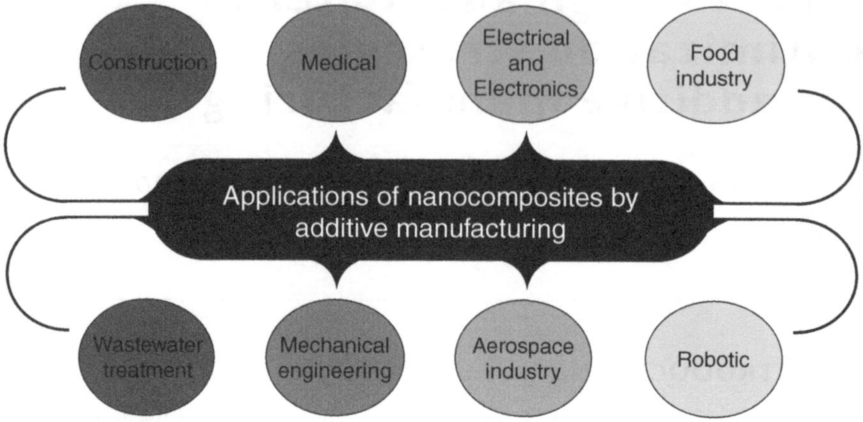

Figure 5.1 Application.

the prosthetic leg by 50% from the original model [2] (see Figure 5.2 for details). The development of 3D printing using machine learning (ML) has been applied to some applications in recent years to optimize the 3D printing performance and processes. For example, Castro et al. [3] have developed an ML model that successfully predicted key aspects of 3D printing formulation and in vitro dissolution properties in drug. It has been found that ML can provide a high level of prediction accuracy, especially in predicting the hot melt extrusion processing temperature, filament aspect, printing temperature, and printability. Also have applied hierarchical machine learning (HML) for the optimization of silicone elastomer 3D printing, in which they were able to increase the printing speed by twofold using the HML algorithm [4]. Recently, Zolfagharian et al. [5] created an ML model to predict the bending angle of a 4D-printed soft pneumatic actuator (SPA) and investigate the effect of the input parameters on its bending. However, the development of 4D printing is still behind as the idea of 4D printing is still under development while 3D printing has been widely commercialized throughout the world.

5.1.1 Topology optimization

Topology optimization is a computational method used to design structures that minimize material usage while maintaining the required performance. It creates optimized material distributions within a defined design space. In AM, topology optimization is used to design lightweight, high-strength structures with less material (refer to Figure 5.3). This is particularly useful for aerospace, automotive, and medical devices. When the topology optimization results are complex lattice structures,

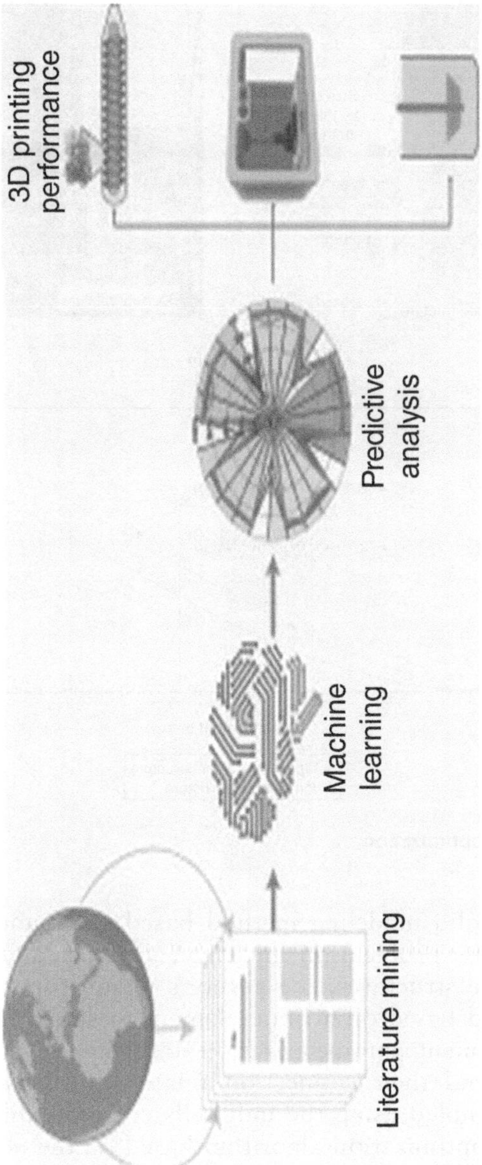

Figure 5.2 Study.

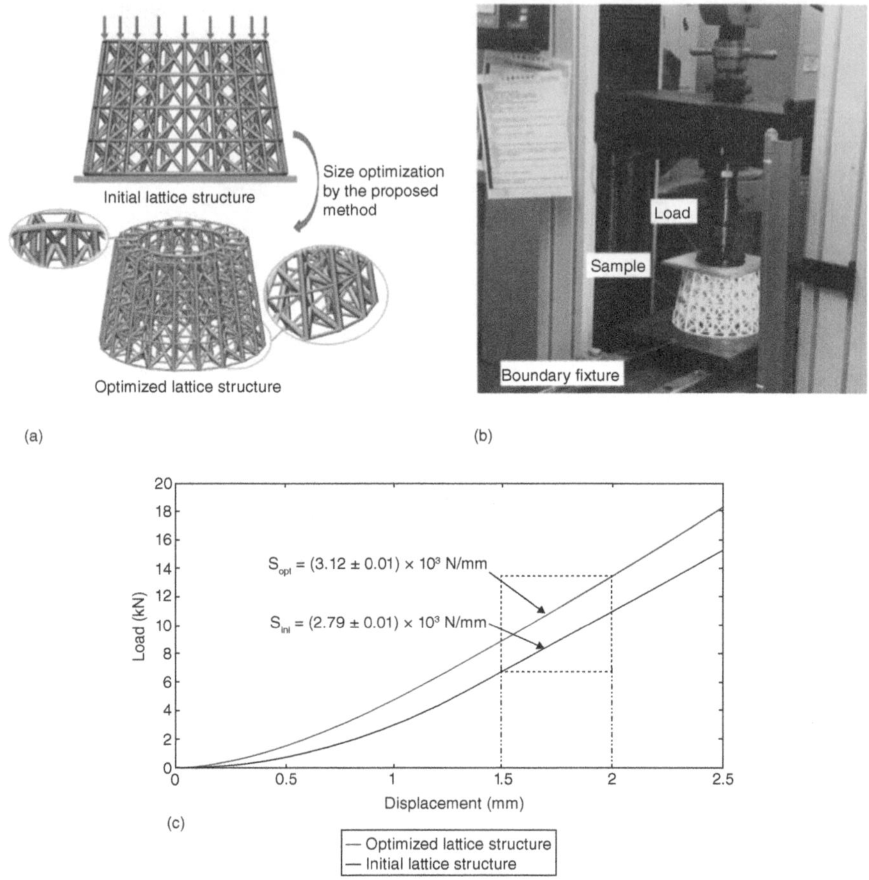

Figure 5.3 Topology optimization.

the traditional solid modeling method based on boundary representation may encounter problems in efficiency, accuracy, and robustness. For large-scale lattice structures, the author's team proposes a parametric modeling method based on a finite element mesh [6–10]. This method describes the element parameterization according to the element type, element node, and their connection relationships. The initial lattice structure is assembled using the unit cells corresponding to each finite element. A size optimization algorithm based on the moving isosurface threshold (MIST) method is proposed to optimize the lattice structure. Graded and heterogeneous lattice structures can be easily constructed using the proposed modeling method.

5.1.2 Methods and algorithms in topology optimization

Density-based methods: The most common method of topology optimization is the density-based approach, where the design space is divided into a grid, and each cell in the grid is assigned a material density value between 0 (no material) and 1 (full material). The optimization algorithm iteratively adjusts the density values to minimize the objective function while satisfying the constraints. *Level set methods:* These methods represent the material boundaries as a level set function and use this function to evolve the shape of the structure during optimization [11–13]. This approach is useful for problems where the design evolves into complex, non-linear shapes.

Evolutionary structural optimization (ESO): ESO is an iterative method where the least efficient elements (in terms of material usage or strength) are progressively removed from the structure. It is a more intuitive approach compared to density-based methods and is especially effective for creating complex, organic shapes. The flowchart outlining the 3D decomposition optimization methodology is shown in Figure 5.1. A single penalty factor of 3 and sensitivity filtering [7] with a filter radius of 2.5 times the average element size are used to obtain the topology-optimized structure, as illustrated in Figure 5.4a. This figure displays only elements with a density greater than 0.5. The 3D decomposition optimization begins by importing the optimal elemental density values, as shown in Figure 5.4b.

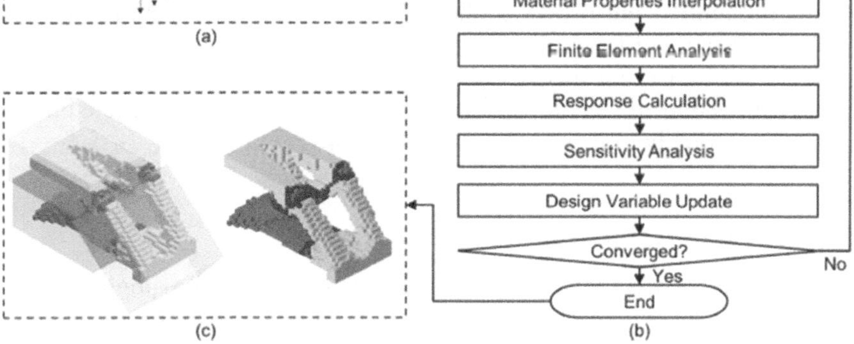

Figure 5.4 Structures.

5.2 GENERATIVE DESIGN

Generative design is often applied in conjunction with AM due to AM's ability to produce highly complex and intricate geometries that are typically not feasible with traditional manufacturing processes [14–16]. The approach enables designers and engineers to create optimized and innovative solutions that would be difficult or impossible to achieve through traditional design methods. Generative design is an algorithmic design process that uses artificial intelligence (AI) to generate multiple design alternatives based on specific goals and constraints. In AM, generative design helps create geometrically complex, efficient parts that would be difficult or impossible to produce with traditional manufacturing methods. Generative design is a powerful tool allowing to fully exploit AM potential. Topology Optimization, already established approach, requires an initial shape where material is removed until the desired result is reached. Generative design presents a more complex approach, receiving several inputs such as objective, constraints, material and manufacturing technique used; it proposes a series of possible outcomes. Designer's role moves from creating the component shape from scratch to choosing the best fitting option for the specific application [17, 18]. The AI tests each potential design against the defined parameters to find the best solution. This approach can result in innovative designs, efficient use of materials, and significant time savings, as it automates much of the trial-and-error work traditionally done by human designers.

Generative design allows manufacturers to input design goals, constraints, and parameters, and then automatically generates a wide range of design alternatives. This feature significantly accelerates the design process, enabling manufacturers to explore more innovative and efficient designs in a fraction of the time it would take manually [19–22]. Generative design optimizes the use of materials by creating lightweight yet strong structures. This is particularly beneficial for reducing material costs and improving sustainability by minimizing waste and reducing the overall carbon footprint of the manufacturing process. The ability to specify performance criteria, such as strength, stiffness, and flow properties, allows generative design to produce solutions that meet or exceed the required performance standards. This ensures that the resulting designs are not only manufacturable but also perform optimally under real-world conditions.

5.3 MULTI-MATERIAL PRINTING

Multi-material AM involves using different materials in a single print job to create parts with varied properties in different regions of the part. This strategy enables more functional parts by combining materials with distinct properties such as flexibility, conductivity, and heat resistance in one component. Extensive research on nature-inspired cellular metamaterials has globally inspired innovations using single material and limited multifunctionality. AM of intricate geometries using multi-materials provides additional functionality, environmental adaptation, and improved mechanical properties [23–26]. Recently, several studies have been conducted on multi-material additive manufacturing (MMAM) technologies, including multi-materials, methodologies, design, and optimization. However, in the past 6 years, very few or no systematic and complete reviews have been conducted in this research domain. This review intends to comprehensively summarize MMAM systems and the working principles of their fundamental processes. Herein, the multi-material combinations and their design, modeling, and analysis strategies have been reviewed systematically. In particular, the focus is on applications and opportunities for using MMAM for several industries and postprocessing MMAM fabricated parts. Furthermore, this review identified the limitations and challenges of existing software packages, MMAM processes, materials, and joining mechanisms, especially at the multi-material interfaces [2]. AM has experienced exponential growth over the past two decades and now stands on the cusp of a transformative paradigm shift into the realm of multi-functional component manufacturing, known as MMAM. While progress in MMAM has been more gradual compared to single-material AM, significant strides have been made in exploring the scientific and technological possibilities of this emerging field. Researchers have conducted feasibility studies and investigated various processes for multi-material deposition, encompassing polymeric, metallic, and bio-materials. To facilitate further advancements, this review paper addresses the pressing need for a consolidated document on MMAM that can serve as a comprehensive guide to the state of the art. Previous reviews have tended to focus on specific processes or materials, overlooking the overall picture of MMAM. Thus, this pioneering review endeavors to synthesize the collective knowledge and provide a holistic understanding of the multiplicity of materials and multiscale processes employed in MMAM. The review commences with an analysis of the implications of multiplicity, delving into its advantages, applications, challenges, and issues. Subsequently, it offers a detailed examination of MMAM with respect to processes, materials, capabilities, scales, and structural aspects. Seven standard AM processes and hybrid AM processes are thoroughly scrutinized in the context of their adaptation for MMAM, accompanied by specific examples, merits, and demerits. The scope of the review encompasses material combinations in polymers, composites, metals–ceramics, metal alloys, and bio-materials. Furthermore, it explores MMAM's capabilities in fabricating bi-metallic structures and functionally (FGMs)/compositionally graded materials, providing insights into various scale and structural aspects [25].

5.4 MATERIAL EFFICIENCY AND WASTE REDUCTION

This strategy focuses on reducing material waste during the printing process by optimizing the design, material selection, and print parameters. Efficient material usage helps minimize costs and environmental impact while ensuring high-quality products. This can be done by optimizing the design to use less material, or by using advanced slicing and printing techniques that reduce waste. Reducing material waste means greater resource efficiency, less pollution, and more profits. Each dollar saved on raw materials costs goes straight to the bottom line. Before you can eliminate raw material waste, you need to be able to identify it. To do this, every aspect of the production process should be addressed and tracked. "The lifecycle flow of materials (e.g., end-use material efficiency improvement and cascading through reuse, recycling, and recovery) and their storage in the economy (stockpiling) are not well understood, and as a consequence, important options for efficiency improvements might be overlooked as attention is focused instead on energy efficiency in materials production." Material efficiency is the practice of using materials and processes to reduce waste, increase output, and use materials more efficiently (Figure 5.5). Waste reduction is the practice of reducing or preventing waste generation.

5.5 FUNCTIONALLY GRADED MATERIALS

FGMs are materials where properties (such as stiffness, thermal conductivity, or density) vary gradually over the material's volume. AM enables the creation of functionally graded parts by allowing for a seamless transition of material properties in a single print job. This is particularly valuable in industries such as aerospace, where components require varying properties in different regions of the part.

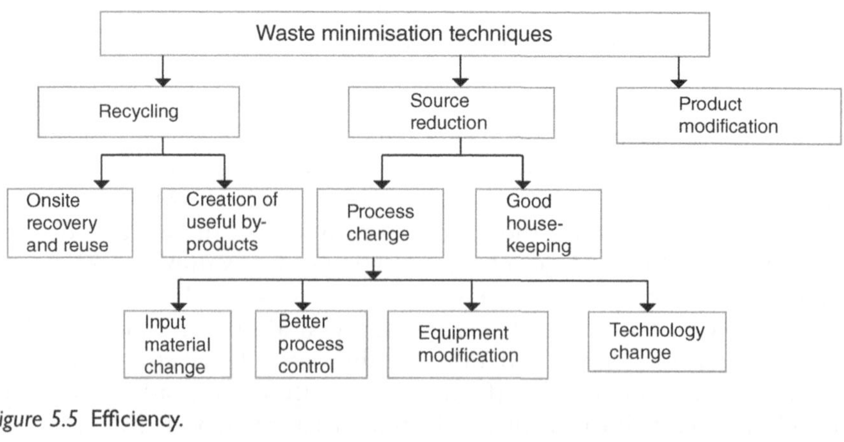

Figure 5.5 Efficiency.

5.5.1 Key characteristics of FGMs

Gradual variation in properties: Unlike traditional composites that have discrete layers, FGMs exhibit a continuous variation in properties. This means the material's characteristics change smoothly across its volume or along its thickness, allowing it to better meet different performance requirements. *Multi-material composition:* FGMs are typically made by blending two or more materials. These materials are combined in such a way that their proportion changes gradually from one surface to another or throughout the structure. *Tailored performance:* FGMs are designed to optimize a specific performance, such as enhancing thermal, mechanical, or electrical properties. For example, they can provide high thermal resistance on one side and high mechanical strength on the other.

5.6 APPLICATIONS

5.6.1 Agriculture

Seed planting and precision farming: AM enables the creation of custom-designed tools for planting, monitoring, and maintaining crops. For example, 3D-printed drones can assist in precision farming by spraying pesticides or fertilizers with minimal waste.

3D printing is the process of layer-by-layer building 3D objects from digital models. Owing to this technology, complex, personalized, and very precise components can now be produced, and this has transformed a number of industries, including aerospace, automotive, and healthcare sectors. The potential of 3D printing has expanded into the food processing and agriculture sectors in recent years, offering to improve efficiency, sustainability, and product quality while addressing a number of issues these sectors encounter [27, 28]. Figure 5.6 showcases the integration of 3D printing, Internet of Things (IoT), and AI technologies in precision agriculture systems, highlighting the connections between custom sensors, IoT networks, and AI-driven data analysis platforms.

5.6.1.1 Custom tools and equipment for seed planting

Farmers can use 3D printing to create customized seed planters that are tailored to specific crop types or field conditions. These planters can be designed for precise depth and spacing, ensuring optimal seed placement for better growth rates and yield. 3D-printed modular tools allow farmers to easily customize and adapt planting systems for different types of terrain and crops. These modular components can be adapted to different types of seeds, planting conditions, and environmental factors. AM allows the design of planting tools with complex geometries that would be difficult

Figure 5.6 Integration.

or impossible to produce with traditional methods. This could include tools with specialized tips, unique shapes for specific crops, or components designed to reduce soil compaction, promoting healthier soil. 3D printing can help design more precise seed-dispensing mechanisms. For example, 3D-printed dispensers can ensure that each seed is dropped in the perfect location, minimizing waste and ensuring that seeds are spaced evenly to maximize growth potential.

5.6.1.2 Drones for seed planting

Drones are increasingly being used for seed planting, especially in hard-to-reach areas or large fields. AM can be used to create lightweight, durable drone components that are optimized for carrying and planting seeds. These drones can be equipped with 3D-printed seed pods or containers, enabling them to dispense seeds more accurately over large areas. 3D-printed drones can be designed to release the right amount of seeds at precise locations, ensuring an even distribution over fields. This reduces seed wastage and optimizes the growth conditions for each plant.

5.6.1.3 Custom fertilizer and pesticide applicators

3D printing can help design specialized nozzles and applicators for fertilizer or pesticide applications that can work in conjunction with seed planting. These can be optimized for specific crops and soil conditions, ensuring that the correct amount of fertilizer or pesticide is applied exactly where needed. With precision tools, farmers can reduce the amount of chemicals used, minimizing waste and the environmental impact. By applying fertilizers or pesticides only where needed, 3D-printed applicators help enhance sustainability in farming.

5.6.1.4 *Prototyping and innovation in agricultural equipment*

Using 3D printing, agricultural companies can prototype and test new seed planting tools much faster than with traditional manufacturing methods. This quick iteration allows for more innovation in farming practices, leading to better, more efficient equipment. Farmers can work with designers or manufacturers to create low-cost, customized tools for their specific needs, whether for small-scale family farms or large commercial operations. This helps farmers adopt the latest technologies without a significant upfront investment. AM allows the creation of biodegradable tools or seed pods that can be used for planting, reducing waste and environmental impact. For example, biodegradable pods that contain seeds and nutrients could be 3D-printed and planted directly into the soil, decomposing as the plants grow. By using 3D printing to create tools that are precisely designed for their specific function, farmers can reduce material waste, as the components are produced with only the exact material required. AM can be used to create custom sensors that can be placed in the soil to monitor moisture, temperature, or nutrient levels. These sensors can help farmers better understand the conditions of their fields and optimize seed planting accordingly. Specialized 3D-printed tools could be used to break up compacted soil or create channels for water infiltration, which is important for seed germination and plant growth. Manufacturers can prototype and customize machinery components such as nozzles or even specialized tractors with specific needs, improving efficiency and reducing operational costs. Using 3D printing for creating biodegradable or organic agricultural tools and containers helps reduce plastic waste, contributing to sustainable farming practices.

5.6.1.5 *Current existing problems of AM technology*

Agriculture and human food safety are closely related, in the use of AM materials used in agricultural production, and we first consider safety issues. The safety problem of AM technology is mainly related to the harm that some printing materials cause to the environment and the human body [29–32]. Taking the most commonly used photocuring materials as an example, the volatile organic compounds they contain will cause certain damage to the environment and human body. The sustainability benefits of 3D printing technology can also play an important role in agricultural production. By redesigning products, minimizing material waste, extending product life, and improving cost efficiency, 3D printing opens up new possibilities for agricultural production [33–36]. Using the design freedom of 3D printing, agricultural producers can create more complex structures and optimized components, thereby improving production efficiency and product quality [37]. Compared to traditional manufacturing technologies, 3D printing has a lower impact on the environment because it reduces energy consumption and material waste. In addition, the use of 3D printing technology can shorten the design-to-production cycle, enabling a more cost-effective production process. At the same time, 3D printing provides a risk-free working environment that helps safeguard the health of agricultural producers. By using 3D printing technology in agricultural production, it can improve the efficiency of material and energy utilization, reduce energy consumption, reduce the negative impact on the environment, and promote the development of agricultural production in a more sustainable direction.

5.6.2 Aviation

AM is extensively used in the aerospace industry to produce lightweight, durable parts that reduce overall aircraft weight, improving fuel efficiency and lowering emissions. Complex geometries that would be difficult to achieve with traditional methods are easily produced via AM. Aircraft manufacturers and maintenance crews use 3D printing to create spare parts for older models or for specific aircraft with low-volume needs. This reduces the need to keep large inventories and lowers downtime for maintenance. 3D printing enables the production of custom components that can be tailored to specific requirements, such as customized cockpit designs, seat structures, or even complex engine parts.

5.6.2.1 Critical attributes and significant applications of AM in the aircraft industry

The ability to convert raw materials into complex 3D forms without the need for elaborate tooling makes AM particularly interesting in the aircraft industry, considering the inventory and supply chain constraints and the need for just-in-time manufacturing. Further, the point-by-point consolidation allows for more complex designs and possible material and performance optimizations, leading to lighter weight and integrated part designs and sustainable performance attributes. Consequently, the leaders of the aircraft industry have been exploring the possible use of AM to produce aircraft parts including various hinges, brackets, interior components, and even the lightweight fuselage and airframe designs, targeting better fuel efficiencies. The actual applications as identified from the literature are as varied as engine components such as turbine blades with internal cooling channels, fuel nozzles and compressors, and integrated piping systems [38, 39].

One of the most significant advantages of AM in aviation is the ability to create lightweight, complex geometries that reduce the overall weight of aircraft. This results in better fuel efficiency, which is crucial for both commercial and military aircraft. Parts that traditionally required multiple components can now be consolidated into a single 3D-printed part, minimizing material waste and weight. AM allows for the production of parts with intricate internal structures that would be impossible or cost-prohibitive to produce using traditional manufacturing methods. For instance, lattice structures, which offer strength while minimizing weight, can be printed directly into aircraft components like brackets, engine parts, and frames. AM is particularly valuable in the aerospace industry for creating spare parts on-demand. Aircraft maintenance often requires parts that are no longer in production or are difficult to source (Figure 5.7). By using 3D printing, airlines and maintenance crews can produce these parts on-site, reducing the need for expensive and time-consuming supply chain logistics. This capability helps reduce aircraft downtime, ensuring faster repairs and lowering operational costs. 3D printing enables parts to be printed in remote locations, including at airports or maintenance facilities. This decentralization can lead to faster turnaround times for repairs, especially for critical parts that are rarely needed but vital when required. AM enables the design and production of highly customized seating arrangements, cabin layouts, and other interior components for both commercial airlines and private jet manufacturers. For example, passengers' comfort can be enhanced by creating personalized

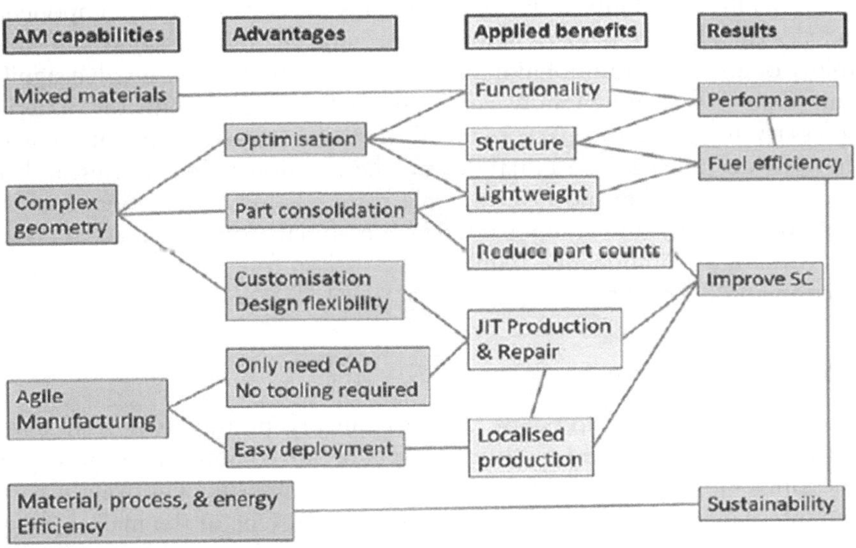

Figure 5.7 AM.

seat designs, or the aircraft's interior could be optimized for unique customer needs, such as luxury or medical transport. AM is used to produce customized cockpit components, including ergonomic panels, handles, or buttons, which are designed for optimal pilot comfort and functionality. These parts can be tailored to the specific needs of an individual pilot or aircraft type.

5.6.2.2 Material advancements for aerospace applications

AM allows the use of specialized materials such as titanium, high-strength aluminum alloys, and composites that are critical for aviation applications. These materials are not only lightweight and strong but also resistant to high temperatures and corrosion, making them ideal for use in aerospace components. The use of metal 3D printing in aviation has grown rapidly, allowing for the production of robust, functional parts for critical systems. On the other hand, polymers, particularly high-performance thermoplastics, are being utilized for non-structural parts like cabin elements and insulation materials. Although 3D printing offers many advantages, the aviation industry remains heavily regulated. However, with advancements in certification processes, 3D-printed parts are increasingly being approved for use in commercial aviation. Companies are working closely with aviation regulators (such as the Federal Aviation Administration (FAA) and European Union Aviation Safety Agency (EASA)) to ensure that 3D-printed components meet stringent safety and performance standards. Manufacturers can use AM to create parts with a higher degree of precision, which can enhance safety. Additionally, the ability to produce complex parts with fewer components reduces the chances of part failure, improving the overall safety and reliability of aircraft. Traditional manufacturing methods often involve significant material waste. AM, however, is more efficient, using only the material necessary to create the part, reducing scrap and waste. This aligns with the growing push for sustainability in the aviation industry. Research is underway to develop more sustainable materials for 3D printing, such as bio-based plastics and composites. These materials could be used to create non-structural components, reducing the carbon footprint of aircraft production and operation.

5.6.3 Automotive

Automotive manufacturers use AM to quickly prototype new designs, allowing them to test and modify concepts before committing to mass production. Complex geometries and innovative designs, like lightweight parts and internal structures, can be created efficiently. One of the main benefits of AM in prototyping is the ability to rapidly produce multiple iterations of a design. Traditional methods like injection molding or CNC machining may take days or even weeks to set up, while 3D printing can produce parts in hours, allowing designers to quickly test and refine their ideas. This significantly accelerates the product development timeline, enabling faster go-to-market times. AM eliminates many of the steps in traditional manufacturing, such as tooling and mold creation, which are time-consuming and costly. Designers can produce prototypes directly from digital models, reducing the lead time for testing, validation, and modification of designs. Just like in aviation, the automotive industry uses AM to produce spare parts quickly and affordably. Custom parts or accessories, such as personalized interiors or unique modifications, can be produced based on customer demand.

In 2010, Ilardo et al. [28] designed and manufactured an intake system for a 600 cc Formula Society of Automotive Engineers engine. In this work, FDM was used to create an intake system, which consisted of a plenum, plenum elbow, and cylinder runners, as shown in Figure 5.3. The use of FDM allowed geometric design freedom, and it could create a unique intake geometry featuring a tapered plenum and tapered runners, generating reduced weight (22% decrease) for the system and enhanced performance.

AM allows designers to create geometries and internal structures that are not possible with traditional manufacturing methods. For example, lattice structures, organic shapes, and complex, multi-material parts can be printed with relative ease. This design freedom enables the creation of more optimized, lightweight, and functional parts, contributing to product innovation. AM offers unparalleled flexibility for customization, allowing products to be tailored to specific requirements. Whether it is a personalized consumer product, a custom medical device, or a bespoke automotive component, AM can create designs that precisely fit individual needs, improving both functionality and customer satisfaction. 3D printing is used to produce jigs, fixtures, and tools, which help streamline production processes, improve accuracy, and reduce lead times, Figure 5.8 shows the example of AM.

5.6.4 Medical

AM is a game changer in creating custom prosthetics and orthotics that are tailored to individual patients' needs, improving comfort and functionality. It also enables more affordable, on-demand production of these devices. One of the most significant advantages of AM is the ability to produce prosthetic and orthotic devices that are specifically designed to fit the unique anatomy of each patient. Traditional manufacturing processes often require a compromise between functionality and comfort. However, AM allows for highly customized designs that perfectly match the patient's measurements, providing a better fit and greater comfort. AM allows for prosthetics and orthotics to be personalized in terms of appearance, color, texture, and even aesthetics. This helps patients feel more confident and comfortable with

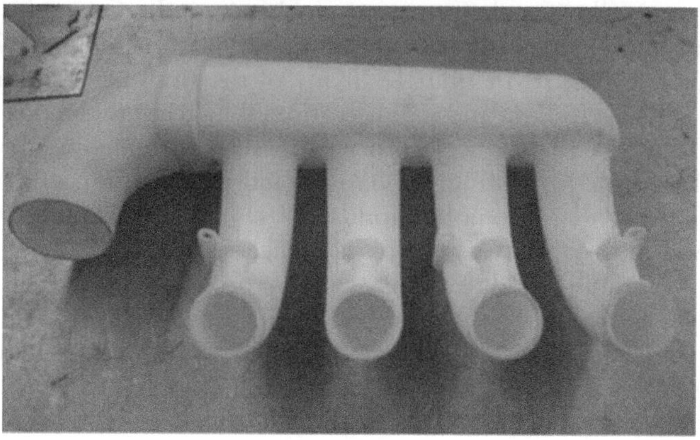

Figure 5.8 AM example.

their devices, reducing the psychological impact of using medical aids. For example, prosthetic limbs can be printed with custom patterns or designs to reflect a patient's personality or preferences.

Surgeons use 3D-printed models of patients' anatomical features to practice surgeries, improving precision and reducing the risk of complications. Custom implants for joint replacements, spinal surgery, and cranial repairs are also produced using AM. AM allows for the creation of highly accurate, patient-specific 3D models of organs, bones, or other anatomical features based on medical imaging data (such as CT or MRI scans). These models provide surgeons with a physical representation of a patient's unique anatomy, which is invaluable for planning complex procedures. Surgeons can examine the model, simulate surgery, and identify potential challenges or complications before performing the actual procedure (see Figure 5.9). In complex surgeries, such as spinal or cardiovascular surgery, AM allows for detailed simulation of the procedure. Surgeons can rehearse the surgery on the 3D-printed model, improving their precision during the actual operation. This preoperative practice reduces the likelihood of errors and increases confidence, leading to better patient outcomes.

Biomedical implants manufactured by AM offer an opportunity to obtain more patient specific, personalized parts with quicker response, a lower inventory level, and reduced delivery costs compared to biomedical implants manufactured by conventional manufacturing, usually accessible to suppliers from the immediate area. Even in the supply chain, it found advantages when compared with traditional methods [41]. Bioprinting is a revolutionary application of AM technology that involves the use of 3D printing to create biological tissues and organs. It utilizes bioinks made from living cells, bio-materials, and growth factors to print functional tissue

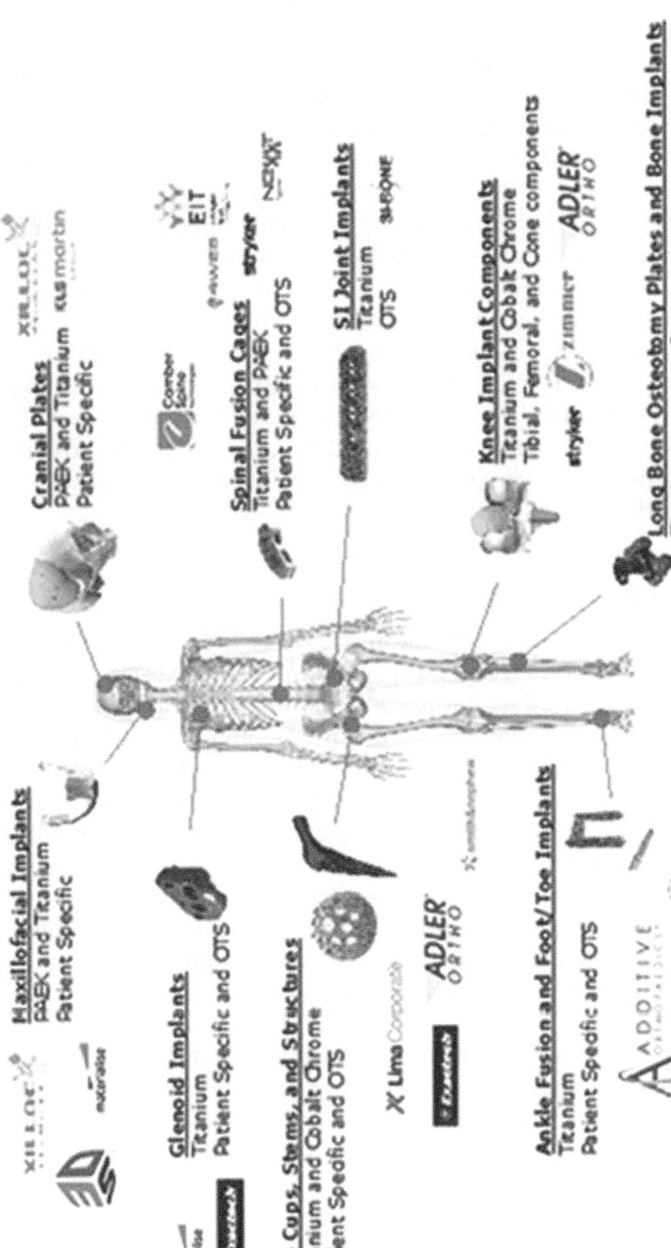

Figure 5.9 Surgery.

structures for research, medical, and therapeutic applications. This field of research has the potential to transform medicine, drug development, and organ transplantation, paving the way for more personalized, sustainable, and effective treatments. Although still in its experimental stage, bioprinting aims to print human tissues and organs. Researchers are using AM to print cellular structures that could eventually lead to organ regeneration. One of the ultimate goals of bioprinting is the creation of functional organs such as kidneys, livers, hearts, or lungs for transplant. While organ transplantation is currently limited by the availability of donor organs, bioprinting holds the potential to solve this problem by creating organs from the patient's own cells, thus eliminating the risk of rejection. Researchers are already printing kidney tissue, heart valves, and small functional organ units. Before full organs can be printed, bioprinting allows for the creation of organ-like models for research. For instance, scientists have successfully printed functional models of human liver and kidney tissues, which can be used for drug testing and disease modeling. These models can simulate the structure and function of a human organ, providing an invaluable tool for studying organ function and disease progression.

5.6.5 Food processing

3D printers for food allow the customization of food shapes, textures, and designs. This technology is especially popular for making chocolate, candies, and intricate sugar-based sculptures. AM in the food industry is being used to produce personalized nutrition. Food printers can create customized meals or supplements with specific nutritional content based on individual health requirements or preferences. AM enables the efficient use of ingredients, reducing food waste. For example, food can be printed with minimal leftover scraps, and ingredients can be precisely measured. Food printers can also work with plant-based or lab-grown proteins to create novel food products, such as meat substitutes or plant-based delicacies, offering new possibilities for sustainable food sources. Apart from carrying out manual operations, autonomous systems that possess 3D-printed sensors have the ability to consistently track the condition of the soil and crops, offering crucial information for targeted farming (more details in Figure 5.10). Drones and robots that use these sensors can monitor and collect data in real time. Drones that are fitted with sensors for measuring soil moisture and 3D-printed multispectral cameras fly over fields and evaluate the health of the crops and soil [40–43].

5.6.6 Sports

AM allows the production of customized sports equipment, such as footwear, helmets, and protective gear, tailored to the specific needs of an athlete.

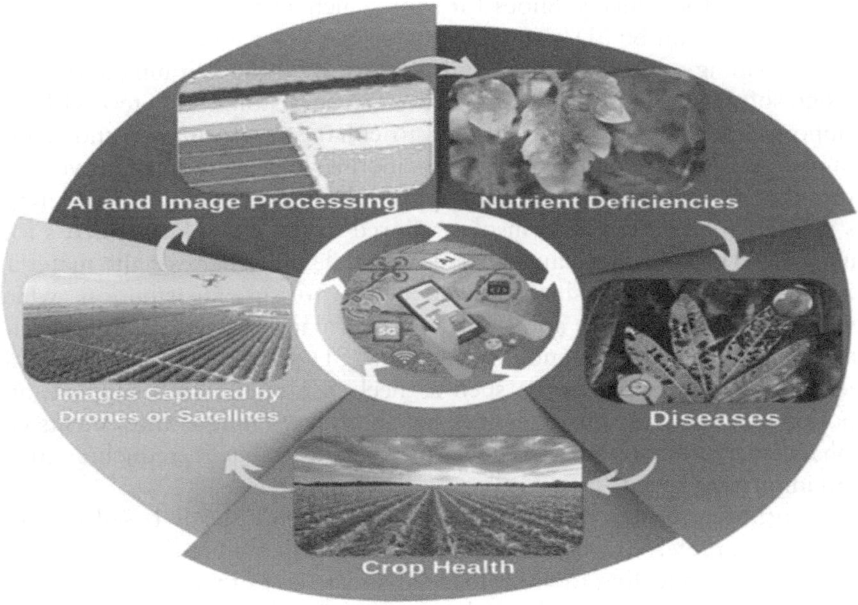

Figure 5.10 Sensors.

These products offer better fit, enhanced performance, and improved safety. In sports such as football, hockey, and cycling, helmets are essential for protecting athletes from head injuries. Traditional helmets are generally produced in standard sizes, but 3D printing allows for the production of custom-fitted helmets based on the individual athlete's head shape. By scanning the athlete's head and using 3D modeling software, manufacturers can create helmets that fit perfectly, offering better comfort and improved protection. Additionally, the design flexibility of 3D printing allows for the integration of advanced shock-absorbing materials to enhance safety. 3D printing can create protective pads and guards that are designed specifically for an athlete's body. Using custom scans or measurements, 3D printers can produce lightweight, high-performance guards that offer enhanced mobility and protection. These pieces can be designed with specialized features such as ventilation holes for breathability, reinforced areas for protection, and ergonomic shapes for better fit and comfort. Custom mouthguards are essential in sports like boxing, rugby, and basketball to protect teeth and reduce the risk of concussions. Using 3D scanning technology, athletes can have mouthguards made to fit the precise contours of their mouth, improving both comfort and protection. These mouthguards can also be designed with impact-absorbing materials to provide better protection during high-impact sports. 3D printing is revolutionizing the design and production of custom sports shoes for athletes. Shoes for sports such as running, soccer, basketball, and golf can be 3D printed to match the exact foot shape of the athlete, providing optimal comfort and support. By using scanning and digital design software, 3D printing allows shoes to be customized in terms of arch support, sole structure, and padding to enhance performance and reduce injury risk. Athletes often require specific insole designs to improve comfort, support, and alignment during physical activity. Using 3D scanning, it is possible to create custom insoles tailored to the individual athlete's foot and gait. These insoles can be made from flexible, lightweight materials that offer increased cushioning and better pressure distribution, which can help prevent injuries and improve performance over time. For athletes who participate in high-endurance sports, 3D-printed running shoes can be tailored for specific needs such as heel cushioning, arch support, and motion control. Customizing shoes for an athlete's unique movements can enhance performance and provide a more comfortable fit, reducing fatigue and improving speed.

3D printing allows golfers to design custom club heads based on their swing style, preferred weight distribution, and grip preference. Using digital scanning and modeling tools, golfers can create clubs that provide the right balance, aerodynamics, and weight to maximize their performance on the course. Custom-made drivers, putters, and irons can lead to more accurate shots and better overall performance. In addition to club heads, 3D printing can also be used to create custom shafts for golf clubs. By customizing the flexibility, weight, and length of the shaft, golfers can fine-tune their clubs to suit their individual playing style and physical characteristics. This ensures a more comfortable and effective game for players of all skill levels. Athletes can also get a completely personalized fitting for their golf equipment, from the shaft material to the clubface angle, with 3D printing. Custom golf club designs can be optimized for precision, durability, and aesthetic appeal, giving athletes more control over their performance. For cycling enthusiasts and professional cyclists, 3D printing allows for the creation of custom bicycle frames tailored to the athlete's body shape, riding style, and performance goals.

5.6.6.1 Custom sports accessories and equipment

Tennis rackets can be 3D printed with customized frames that match an athlete's preferences for balance, flexibility, and string tension. By analyzing the athlete's grip and swing style, 3D printing allows for the design of a racket that feels more natural and improves performance. Custom rackets can also be created with personalized grip sizes and handle designs, ensuring a better fit for the player's hand and reducing the risk of discomfort or injury during extended play. Skiers and snowboarders can benefit from custom-fitted boots and bindings that are designed for their unique foot shape and riding style. 3D printing allows for the production of highly personalized ski boots that offer superior comfort, control, and warmth. The boots can be printed with the perfect fit, improving performance on the slopes and minimizing the risk of injury. Swimmers can also take advantage of custom 3D-printed goggles, swimsuits, and flippers. Goggles can be made to fit precisely, eliminating water leakage, while swimsuits can be tailored for specific needs such as drag reduction or flexibility. These custom designs can enhance an athlete's performance and comfort in the water.

5.6.6.2 Prototyping and design innovation

AM allows sports equipment manufacturers to create rapid prototypes of new designs. This allows for quick iterations of gear prototypes, enabling designers to test different features, shapes, and materials before committing to large-scale production. Whether it is a new racquet design, shoe prototype, or sports equipment accessory, 3D printing can help innovators bring their ideas to life faster and more cost-effectively. The flexibility of 3D printing also allows for experimentation with new materials to enhance the performance and durability of sports gear. Many 3D-printed sports equipment pieces, such as bicycles, golf clubs, and skateboard decks, can be created with optimal strength-to-weight ratios. This means athletes can use lighter gear without sacrificing strength or durability, improving their agility and speed during sports activities. Athletes use 3D-printed devices for injury recovery, such as custom-designed braces or supports that provide better comfort and support.

5.7 CONCLUSION

Smart design and material optimization are critical for maximizing the efficiency and potential of AM. By leveraging advanced techniques such as topology optimization, generative design, multi-material printing, and FGMs, manufacturers can produce parts that are lighter, stronger, and more cost-effective. As technology continues to evolve, these strategies will play a key role in advancing industries like aerospace, automotive, medical devices, and consumer electronics, driving innovation and sustainability. These strategies enable the creation of highly efficient, cost-effective, and sustainable products. By leveraging advanced design tools such as generative design and topology optimization, engineers can push the boundaries of AM, producing parts with complex geometries that traditional manufacturing methods cannot achieve. Material optimization, which includes the selection of materials based on specific performance requirements, durability, and cost, ensures that AM can be applied to a wide range of industries, from aerospace to healthcare. AM has revolutionized multiple industries by providing more efficient, flexible, and cost-effective ways to produce highly customized and complex products. From reducing waste in food production to enabling cutting-edge medical devices and customizing automotive parts, its role in innovation and sustainability continues to grow across agriculture, aviation, automotive, medical, food processing, and sports sectors. As the technology continues to advance, these applications will likely expand and evolve further, providing new solutions to long-standing challenges.

REFERENCES

1. X. Li, S. Zhang, P. Jiang, M. Nie, D. Kong, Z. Kang, M. Liu, D. Zhu, C. Jiang, Q. Zhang, S. Zu, Z. Zhang. Smart bionic structures: connecting nature and technology through additive manufacturing. *Additive Manufacturing Frontiers*, 2024, 3(2), 200137, ISSN 2950-4317, https://doi.org/10.1016/j.amf.2024.200137
2. M.Z. Cordero, R. Dedic, Z. Jelacic, R. Toshev. Optimizing mechanical design for an additively manufactured prosthetic leg. *Procedia Computer Science*, 2024, 232, 2973–2983, ISSN 1877-0509, https://doi.org/10.1016/j.procs.2024.02.113
3. B.M. Castro, M. Elbadawi, J.J. Ong, T. Pollard, Z. Song, S. Gaisford, G. Pérez, A.W. Basit, P. Cabalar, A. Goyanes. Machine learning predicts 3D printing performance of over 900 drug delivery systems. *Journal of Controlled Release*, 2021, 337, 530–545, ISSN 0168-3659, https://doi.org/10.1016/j.jconrel.2021.07.046
4. D.-G. Shin, T.-H. Kim, D.-E. Kim. Review of 4D printing materials and their properties. *International Journal of Precision Engineering and Manufacturing-Green Technology*, 2017, 4, 349–357.
5. A. Zolfagharian, L. Durran, K. Sangtarash, S.E. Ghasemi, A. Kaynak, Mahdi Bodaghi. 4 – 4D printing modeling via machine learning. In *Additive Manufacturing Materials and Technologies, Smart Materials in Additive Manufacturing*, Elsevier, 2022, 73–102, ISBN 9780323954303, https://doi.org/10.1016/B978-0-323-95430-3.00004-X
6. W. Chen, X. Zheng, S. Liu. Finite-element-mesh based method for modeling and optimization of lattice structures for additive manufacturing. *Materials*, 2018, 11(11), 2073.
7. O. Sigmund. Morphology-based black and white filters for topology optimization. *Structural and Multidisciplinary Optimization*, 2007, 33, 401–424.
8. H. Kim, I.Y. Kim. 3D decomposition optimization of topology-optimized structures considering a build volume constraint for additive manufacturing. *Computer Methods in Applied Mechanics and Engineering*, 432, Part A, 2024, 117357, ISSN 0045-7825, https://doi.org/10.1016/j.cma.2024.117357
9. S.C. Altıparmak, V.A. Yardley, Z. Shi, J. Lin, Challenges in additive manufacturing of high-strength aluminium alloys and current developments in hybrid additive manufacturing. *International Journal of Lightweight*

Materials and Manufacture, 2021, 4(2), 246–261, ISSN 2588-8404, https://doi.org/10.1016/j.ijlmm.2020.12.004
10. I. Polmear, et al. *Light Alloys: Metallurgy of the Light Metals*. Butterworth-Heinemann, 2017.
11. K. Sripriyan, M. Karthigha. A comparative study on the microstructures and mechanical properties of arc welded of HSLA steel. *Journal of Mechanical Science and Technology*, 2024a, 38(12), 6791–6798.
12. K. Adam, D. Zöllner, D. P. Field. 3D microstructural evolution of primary recrystallization and grain growth in cold rolled single-phase aluminum alloys. *Modelling and Simulation in Materials Science and Engineering*, 2018, 26(3), 035011.
13. M. Ma, Z. Wang, X. Zeng. "A comparison on metallurgical behaviors of 316L stainless steel by selective laser melting and laser cladding deposition." *Materials Science and Engineering: A*, 2017, 685, 265–273.
14. T. Heitkamp, K. Hilbig, M. Goutier, S. Girnth, N. Waldt, G. Klawitter, T. Vietor. Characterization of mechanical properties, efficiency, and design considerations for the additive manufacturing of hybrid composites with continuous fibers. *Hybrid Advances*, 2024, 5, 100146, ISSN 2773 207X, https://doi.org/10.1016/j.hybadv.2024.100146
15. C. Huang, M.W. Joosten. 3D printed continuous fibre-reinforced composites: design and characterisation of advanced pseudo-ductile hybrid laminates. *Composites Part A: Applied Science and Manufacturing*, 2021, 146, 106403.
16. K. Wang, et al. Simultaneous reinforcement of both rigidity and energy absorption of polyamide-based composites with hybrid continuous fibers by 3D printing. *Composite Structures*, 2021, 267, 113854.
17. T. Heitkamp, et al. "Continuous fiber-reinforced material extrusion with hybrid composites of carbon and aramid fibers." *Applied Sciences*, 2022, 12(17), 8830.
18. Ali Akmal Zia, et al. Mechanical and energy absorption behaviors of 3D printed continuous carbon/Kevlar hybrid thread reinforced PLA composites. *Composite Structures*, 2023, 303, 116386.
19. S. Li, et al. Investigation on the mechanical properties of 3D printed hybrid continuous fiber-filled composite considering influence of interfaces. *The International Journal of Advanced Manufacturing Technology*, 2022, 123(9), 3147–3158.
20. T. Heitkamp, et al. Experimental and numerical investigation of the mechanical properties of 3D-printed hybrid and non-hybrid composites. *Polymers*, 2023, 15(5), 1164.
21. G. Kretsis. A review of the tensile, compressive, flexural and shear properties of hybrid fibre-reinforced plastics. *Composites*, 1987, 18(1), 13–23.
22. G. R. Chavhan, L. N. Wankhade. Improvement of the mechanical properties of hybrid composites prepared by fibers, fiber-metals, and nano-filler particles – a review. *Materials Today: Proceedings*, 2020, 27, 72–82.
23. H. Schurmann. *Konstruieren mit faser-kunststoff-verbunden*. Vol. 2. Berlin: Springer, 2005.
24. K. Czech, R. Oliwa, D. Krajewski, K. Bulanda, M. Oleksy, G. Budzik, A. Mazurkow. Hybrid polymer composites used in the arms industry: a review. *Materials*, 2021, 14(11), 3047.

25. A. Nazir, O. Gokcekaya, K. M. M. Billah, O. Ertugrul, J. Jiang, J. Sun, S. Hussain. Multi-material additive manufacturing: a systematic review of design, properties, applications, challenges, and 3D printing of materials and cellular metamaterials. *Materials & Design*, 2023, 226, 111661, ISSN 0264-1275, https://doi.org/10.1016/j.matdes.2023.111661
26. A. Verma, A. Kapil, D. Klobčar, A. Sharma. A review on multiplicity in multi-material additive manufacturing: process, capability, scale, and structure. *Materials*, 2023, 16, 5246. www.mdpi.com/1996-1944/16/15/5246
27. Y. Liu, X. Liu, J. Lu, K. Luo, Z. Zhang, H. Lu, H. Zhang, X. Xu, Y. Wang, S. Zhou. Post-treatment technologies for high-speed additive manufacturing: status, challenge and tendency. *Journal of Materials Research and Technology*, 2024, 30, 1057–1082, ISSN 2238-7854, https://doi.org/10.1016/j.jmrt.2024.03.110
28. R. Ilardo, C.B. Williams. Design and manufacture of a formula SAE intake system using fused deposition modeling and fiber-reinforced composite materials. *Rapid Prototyping Journal*, 2010, 16, 174–179.
29. H. Rogers, M. Srivastava. Emerging sustainable supply chain models for 3D food printing. *Sustainability*, 2021, 13(21), 12085.
30. T.D. Ngo, A. Kashani, G. Imbalzano, K.T.Q. Nguyen, D. Hui. Additive manufacturing (3D printing): a review of materials, methods, applications and challenges. *Composites Part B: Engineering*, 2018, 143, 172–196.
31. A. Bhatia, A.K. Sehgal. Additive manufacturing materials, methods and applications: a review. *Materials Today: Proceedings*, 2023, 81, 1060–1067.
32. R.J.M. Wolfs, T.A.M. Salet, N. Roussel. Filament geometry control in extrusion-based additive manufacturing of concrete: the good, the bad and the ugly. *Cement and Concrete Research*, 2021, 150, 106615.
33. R. Chen, H. Yin, I.S. Cole, S. Shen, X. Zhou, Y. Wang, S. Tang. Exposure, assessment and health hazards of particulate matter in metal additive manufacturing: a review. *Chemosphere*, 2020, 259, 127452.
34. A. Jadhav, V.S. Jadhav. A review on 3D printing: an additive manufacturing technology. *Materials Today: Proceedings*. 2022, 62, 2094–2099.
35. C. Maraveas, I.S. Bayer, T. Bartzanas. 4D printing: perspectives for the production of sustainable plastics for agriculture. *Biotechnology Advances*, 2022, 54, 107785.
36. M. Amini, A. Rahmani. Agricultural databases evaluation with machine learning procedure. *Australian Journal of Basic and Applied Science*, 2023, 8, 39–50.
37. K. Sripriyan, M. Karthigha. Influence of granite micro particles on the mechanical, damping and antimicrobial properties of silk–sisal hybrid composites. *Scientific Reports*, 2024b, 14(1), 1–14.
38. B. Heiden, V. Alieksieiev, M. Volk, B. Tonino-Heiden. Framing artificial intelligence (AI) additive manufacturing (AM). *Procedia Computer Science*, 2021, 186, 387–394.
39. M.R. Khosravani, T. Reinicke. On the environmental impacts of 3D printing technology. *Applied Materials Today*, 2020, 20, 100689.
40. N. Guo, M.C. Leu. Additive manufacturing: technology, applications and research needs. *Frontiers in Mechanical Engineering*, 2013, 8, 215–243.

41. T. Catts. GE turns to 3D printers for plane parts. *Bloomberg Businessweek*. 2013.
42. A. Bernier-Vega, et al. Remote data acquisition using UAVs and custom sensor node technology. *Drones*, 2023, 7(6), 340.
43. M. Pant, et al. Additive manufacturing: the significant role in biomedical and aerospace applications. *Indian Journal of Engineering & Materials Sciences*, August 2021, 28, 330–342.

Chapter 6

Experimental investigation of amplitude suppression in PLA cantilevers using magnetorheological fluid and piezoelectric macro-fiber composites under forced vibration

Ramkumar Gunasekaran

6.1 INTRODUCTION

Magnetorheological (MR) fluids exhibit an increase in viscosity with the augmentation of magnetic field strength, attributed to the formation of chain-like structures aligned with the applied magnetic field direction [4]. Previous investigations into MR-based structures have utilized various host materials, such as polyvinyl chloride [1], aluminum [8, 9], and steel [10]. The controllability of natural frequencies has been demonstrated through the manipulation of magnetic field strength, as evidenced by MR fluid analysis in sandwich structures. Moreover, the application of a magnetic field has been shown to suppress vibration amplitudes effectively. Piezoelectric materials exhibit mechanical deformation in response to electric charge stimuli. Extensive studies on the vibration characteristics of piezoelectric materials have explored various geometries [11–16]. The damping ratios of polymers and elastomers typically range from 0.01 to 0.5, with variations stemming from structural disparities tailored for specific applications [5–7].

In this study, polylactic acid (PLA) was selected due to its capacity for fabricating intricate shapes via three-dimensional (3D) printing and its biodegradable properties. External vibrations can compromise the performance of grippers constructed from PLA. Therefore, this study investigates PLA gripper fingers, one with an MR fluid core and another with a piezoelectric macro-fiber composite (MFC)–P2 (contraction) core, for amplitude suppression under forced vibration. Previous research has discussed the development and damping characteristics of PLA grippers with MR fluid [2], PLA grippers with MFC [2], and PLA cantilevers with MR fluid [3]. The PLA–MR fluid gripper specimen, PLA–MFC gripper specimen, and PLA cantilever specimen with MR fluid are depicted in Figures 6.1, 6.2, and 6.3, respectively.

 DOI: 10.1201/9781003569633-6

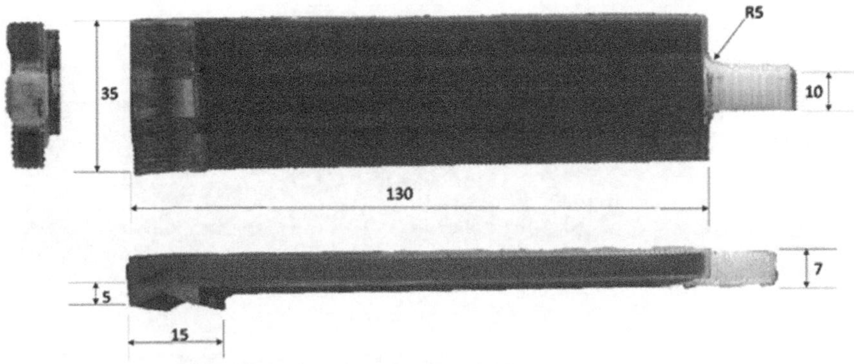

Figure 6.1 The PLA–MR fluid gripper specimen.

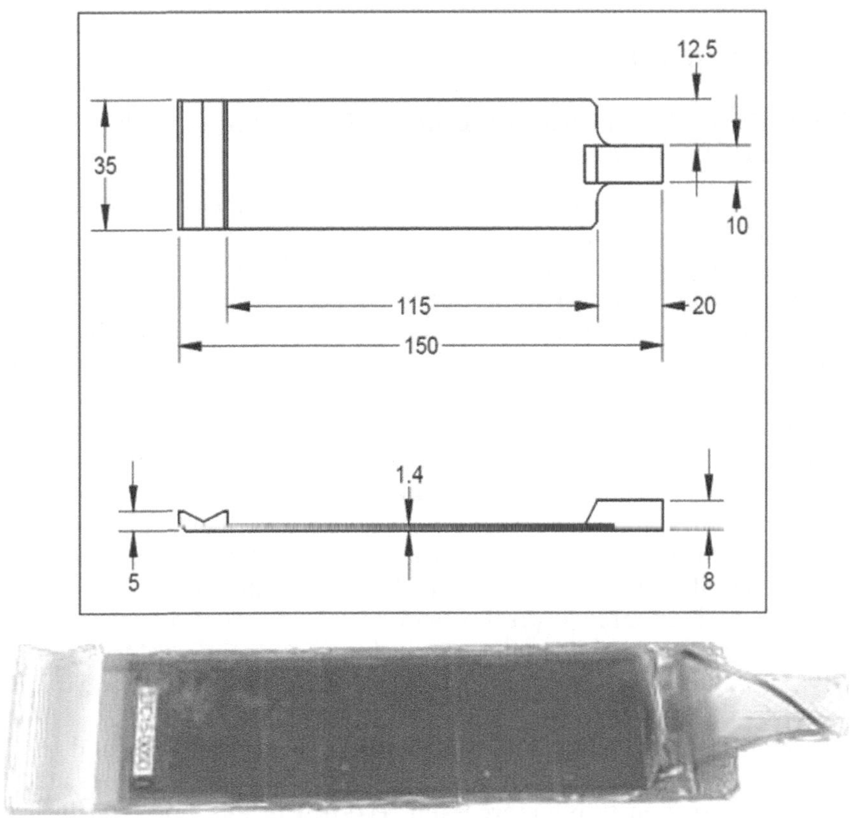

Figure 6.2 PLA–MFC gripper specimen.

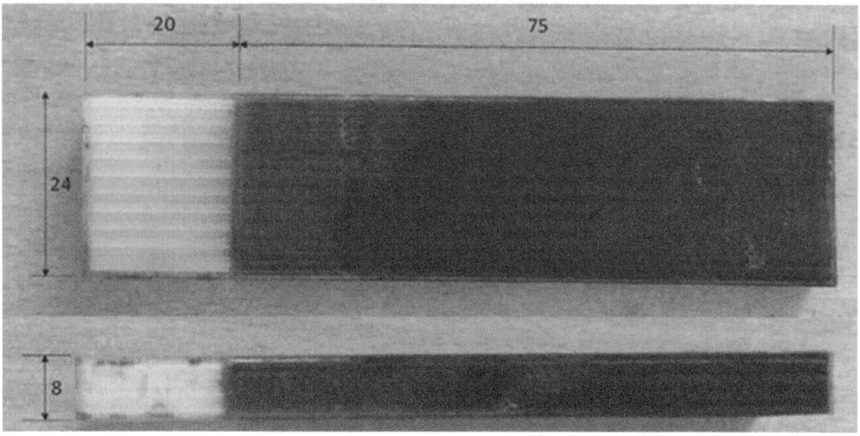

Figure 6.3 PLA cantilever specimen with MR fluid.

6.2 EXPERIMENTAL STUDY

6.2.1 Forced vibration response of specimens with MR fluid core

To investigate the forced vibration response of specimens featuring an MR fluid core, an amplified piezoelectric actuator (APA) is employed as a harmonic exciter. The APA is chosen for its capability to provide precise and controlled vibrations across a wide frequency range. The experimental setup is meticulously designed to ensure accurate and reproducible results. A LabVIEW program is meticulously programmed to generate a signal that is then transmitted to a micro-amplifier. This micro-amplifier is responsible for driving the APA, which in turn induces vibrations in the specimens under investigation. The control over the signal generation process allows for precise modulation of the vibration parameters, such as frequency and amplitude, ensuring consistency throughout the experiment. Data acquisition is facilitated through a National Instruments (NI) Data Acquisition (DAQ) system, specifically the NI9263 module. This module interfaces with the micro-amplifier, enabling real-time monitoring and control of the vibration excitation process. Additionally, an accelerometer, connected to a separate NI9234 module, is strategically placed to sense and record the vibration responses of the specimens.

The experimental procedure involves subjecting the cantilever and gripper specimens to harmonic excitation using the APA. The response of each specimen is meticulously monitored and recorded in the time domain using the accelerometer. To comprehensively investigate the effect of magnetic flux density on the vibration response, the distance between the permanent

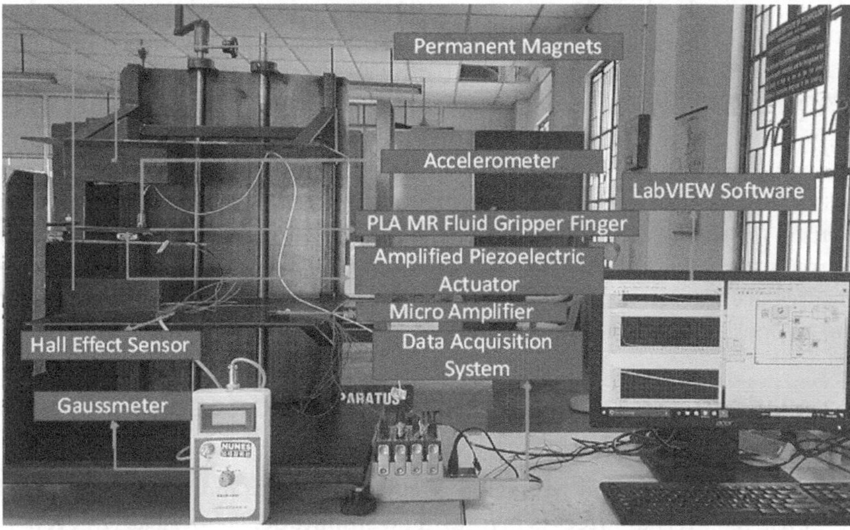

Figure 6.4 Setups.

magnets is systematically varied to produce different magnetic field strengths (0 G, 200 G, 400 G, and 600 G). This meticulous approach ensures that a comprehensive dataset is obtained, allowing for detailed analysis of the forced vibration response of specimens with an MR fluid core. Figure 6.4 provides a visual representation of the sophisticated experimental setup utilized for this purpose.

6.2.2 Forced vibration response of PLA–MFC gripper

This section investigates the forced vibration response analysis of the PLA–MFC gripper. The experimental setup utilizes a sophisticated micro-amplifier featuring two input terminals and two output terminals. The first input terminal receives signals from the LabVIEW program, facilitating the precise actuation of the MFC. Consequently, the generated signal can reach a maximum output voltage of up to 360 V, ensuring meticulous control over the activation of the MFC. Simultaneously, the second input terminal is employed to excite the APA harmonically, based on signals generated within a separate LabVIEW program.

To orchestrate these intricate operations seamlessly, the DAQ NI9263, equipped with eight terminals, is deployed. The first two terminals of the DAQ module are meticulously connected to the MFC and APA, respectively. This configuration enables the simultaneous operation of both programs, thereby ensuring continuous actuation of the MFC and excitation of the gripper throughout the experimental procedure. The LabVIEW program plays a pivotal role in generating signals transmitted to the micro-amplifier via the DAQ NI9263, thereby driving the actuator and MFC in a coordinated manner. Vibration sensing is facilitated by an accelerometer, connected to the DAQ NI9234, which meticulously captures the vibration responses of the PLA–MFC gripper. Subsequently, time-domain plots are generated to analyze the response across different electric field strengths (0 V, 90 V, 180 V, and 270 V). Figure 6.5 provides a visual representation of the experimental setup utilized for the forced vibration analysis of the PLA–MFC gripper. This setup, characterized by its precision and integration of advanced instrumentation, ensures accurate and comprehensive data acquisition for in-depth analysis of the gripper's vibration response under varying electric field conditions.

6.3 THEORETICAL AMPLITUDE UNDER FORCED VIBRATION

In the realm of forced vibration analysis, understanding the amplitude response to external forces is pivotal. The amplitude, denoted as *X*/*F*, reflects the relationship between the applied force and resulting displacement. It can be mathematically expressed as:

$$X / F = \frac{1}{k\sqrt{\left(1 - r^2\right) + \left(2\varepsilon r\right)^2}} \tag{6.1}$$

where k signifies the stiffness of the specimen, a critical parameter influencing its response to external forces [17, 18]. The stiffness (k) is computed using the equation $k = \omega n2m$, where ωn represents the natural frequency and m denotes the mass of the specimen. The total mass of PLA–MR fluid beam, PLA–MR fluid gripper, and PLA–MFC gripper specimens are 0.028 kg, 0.112 kg, and 0.01 kg, respectively.

The theoretical foundation of forced vibration analysis often relies on the equation of motion, a cornerstone in understanding the dynamic behavior of mechanical systems. This equation encapsulates the intricate interplay between external forces, mass, damping, and stiffness within the system. It can be succinctly articulated as:

$$m(d^2x/dt^2) + c(dx/dt) + kx = F0\ sin(\omega t) \tag{6.2}$$

In this equation, the variables are defined as follows:

m: Mass of the system, representing its inertial properties.
x: Displacement of the system from its equilibrium position, a key parameter indicating the extent of vibration.

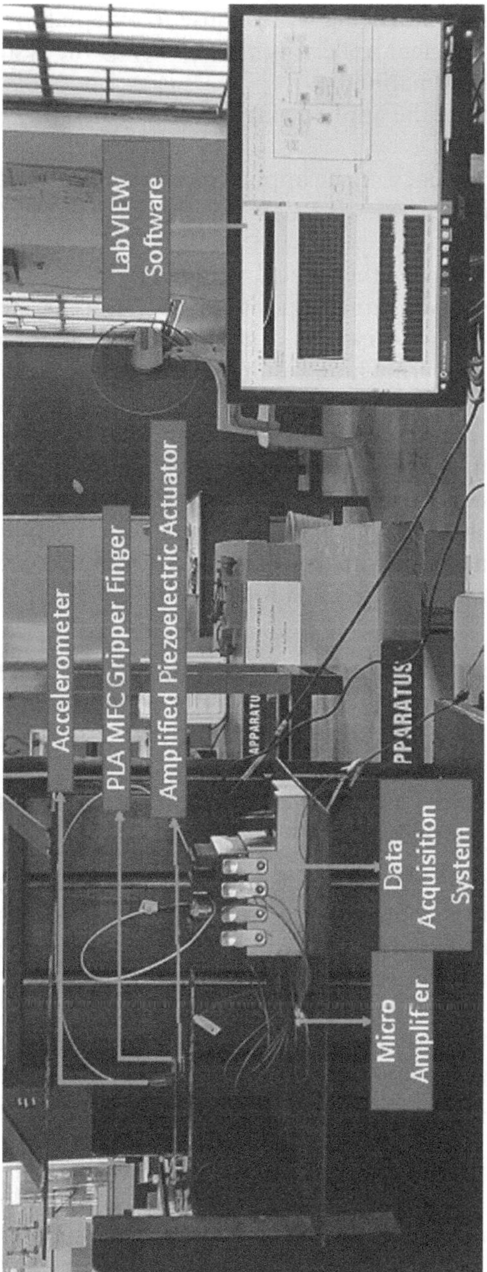

Figure 6.5 PLA–MFC gripper.

t: Time, acting as a temporal reference for dynamic behavior.
c: Damping coefficient, serving as a measure of damping within the system and influencing its response to external forces.
k: Stiffness coefficient, reflecting the rigidity of the system and its resistance to deformation.
$F0$: Amplitude of the applied external force, dictating the magnitude of excitation.
ω: Angular frequency of the applied external force, determining the rate of oscillation.

This comprehensive theoretical framework provides a nuanced understanding of the dynamic response of mechanical systems under forced vibration conditions. It serves as a guiding principle for the analysis and interpretation of experimental results obtained through rigorous testing methodologies.

The schematic diagram depicted in Figure 6.6 intricately illustrates the experimental setup, showcasing vital components such as the accelerometer, PLA–MFC gripper finger, AMA, micro-amplifier, data acquisition system, and LabVIEW software. These components collectively contribute to the precise measurement and analysis of forced vibration responses, underscoring the meticulous approach adopted in this study.

6.4 ROLE OF ARTIFICIAL INTELLIGENCE IN AMPLITUDE SUPPRESSION OF PLA CANTILEVER BEAMS WITH MR FLUID CORE AND ROBOTIC GRIPPER SYSTEM

6.4.1 Introduction to artificial intelligence in structural vibration control

Artificial intelligence (AI) and machine learning (ML) are revolutionizing structural vibration control systems by enabling adaptive, precise, and energy-efficient mechanisms. Traditional vibration suppression techniques rely on passive or semi-active materials and systems that often lack real-time adaptability. AI algorithms address these limitations by dynamically analyzing vibration patterns and enabling intelligent control strategies to optimize damping responses.

In structural vibration control, AI plays a pivotal role in enhancing the performance of smart materials such as MR fluids and MFC. ML models can analyze complex vibration datasets, predict responses under varying loading conditions, and fine-tune control strategies. These capabilities ensure rapid and effective amplitude suppression, improving system efficiency in applications like cantilever beams and robotic gripper systems.

6.4.2 Application of AI in monitoring and data analysis

6.4.2.1 Data collection and preprocessing

Accurate data acquisition is the cornerstone of AI-based vibration control. Using tools like LabVIEW for vibration testing, data on displacement, acceleration, frequency, and external field intensities (magnetic or electric) can be collected in real time. AI algorithms, including signal processing techniques, are employed to preprocess data by filtering noise and anomalies using the Fourier transform and wavelet analysis and clean and normalize datasets for efficient analysis.

6.4.2.2 Feature extraction

Feature extraction focuses on identifying critical parameters that influence vibration behavior. AI techniques such as neural networks and Principal Component Analysis (PCA) enable recognition of dominant vibration frequencies and amplitudes and extraction of significant features such as peak acceleration, damping ratios, and magnetic/electric field intensities. For example, in forced vibration testing of PLA MR fluid beams and grippers, AI can identify patterns in amplitude suppression corresponding to the variation in external fields.

6.5 AI FOR REAL-TIME CONTROL AND ADAPTIVE DAMPING

6.5.1 Predictive modeling for vibration response

ML models, such as regression models, decision trees, and deep learning algorithms, are employed to predict the vibration response of structures under varying magnetic and electric field intensities. These models can forecast amplitude suppression based on input conditions such as magnetic strength (0–600G) and electric fields (0–360V) and enable proactive control strategies to optimize damping.

6.5.2 Adaptive control strategies

AI-driven adaptive control systems utilize Reinforcement Learning (RL) and Adaptive Neuro-Fuzzy Inference Systems (ANFIS) to dynamically adjust system parameters. For instance, RL algorithms continuously monitor feedback from the structure's vibration sensors to adjust magnetic or electric field intensities in real time; ANFIS combines fuzzy logic and neural networks to achieve adaptive tuning of damping forces for optimal amplitude suppression.

6.5.3 Optimization of magnetic field intensity

Optimization algorithms like genetic algorithms (GAs) and particle swarm optimization (PSO) are applied to identify the optimal positioning and strength of magnetic fields for effective suppression and balance power consumption and system performance. For instance, magnetic fields applied to PLA–MR fluid beams can be optimized to achieve amplitude suppression of up to 54.66%, as demonstrated in the experimental results.

6.6 AI IN EXPERIMENTAL DESIGN AND PARAMETER TUNING

6.6.1 Experimental parameter optimization

AI techniques like Bayesian optimization and gradient-based methods are employed to fine-tune critical experimental parameters, including MR fluid viscosity and its response to varying magnetic fields and electric field intensities applied to MFC-based systems. By iteratively testing parameter combinations, AI ensures the most effective amplitude suppression under forced vibration conditions.

6.6.2 Simulations and digital twins

AI enables the development of digital twins—virtual replicas of the physical cantilever beam and robotic gripper systems. These simulation models can predict the system's response to different input conditions (e.g., magnetic field variations) and optimize experimental parameters before physical testing, thus reducing time and resource requirements. For example, simulations of the PLA–MR fluid beam can align closely with experimental results, showing only an average error percentage of 8.1%.

6.7 AI FOR PREDICTIVE MAINTENANCE AND DIAGNOSTICS

6.7.1 Health monitoring and predictive maintenance

AI-powered health monitoring systems leverage predictive maintenance algorithms to detect anomalies or wear in the PLA cantilever beams, MR fluids, or MFC grippers. These systems analyze real-time sensor data to predict component degradation and alert users to potential failures, improving reliability and extending system life. For instance, AI can monitor amplitude variations and identify early signs of wear that might compromise vibration control efficiency.

6.7.1.1 *Diagnostics of system components*

AI algorithms can diagnose faults in the PLA material, MR fluid behavior, or electric field response by identifying deviations from normal vibration patterns. This ensures consistent performance and minimal downtime for structural systems.

6.8 RESULTS AND DISCUSSION

Figures 6.6 and 6.7 depict the forced vibration responses of the PLA–MR fluid cantilever beam and PLA–MR fluid gripper, respectively. Remarkably, there is a substantial suppression in amplitude (acceleration) by 54.66% and 67.09%, respectively, attributed to the increase in magnetic field strength from 0 G to 600 G. Additionally, Figure 6.8 presents the forced vibration response of the PLA–MFC gripper, showcasing a significant suppression in amplitude (acceleration) by 58.64% due to the escalation in the electric field from 0 V to 360 V.

To delve deeper into the analysis, it is imperative to understand the theoretical framework governing forced vibration amplitude. The theoretical forced vibration amplitude in displacement is transformed into acceleration, factoring in the derivative of displacement and the square of the undamped natural frequency in radians per second. Figures 6.9 to 6.11 offer a comparative analysis between theoretical and experimental forced

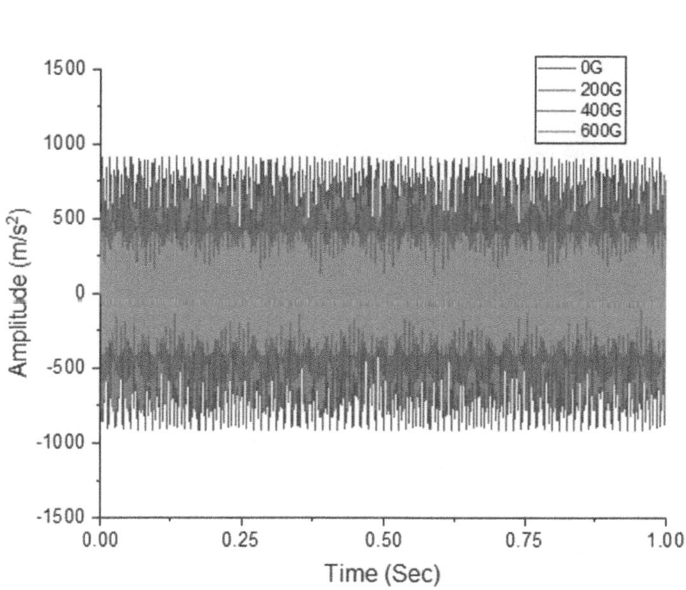

Figure 6.6 The forced vibration responses of the PLA–MR fluid gripper.

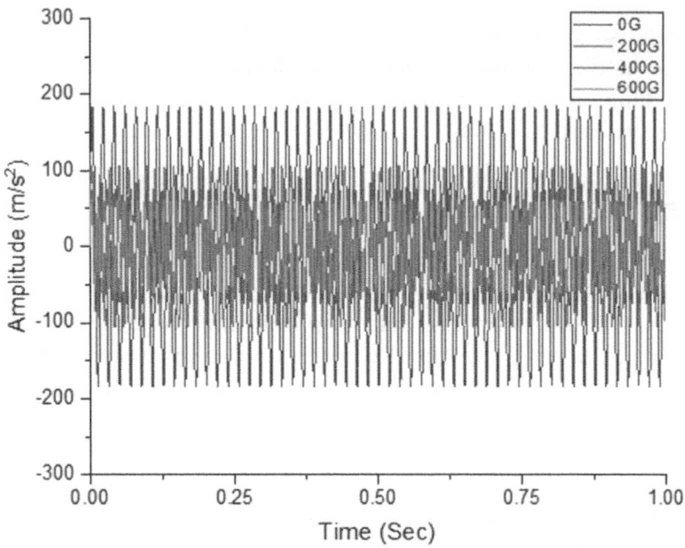

Figure 6.7 The forced vibration responses of the PLA–MR fluid cantilever beam.

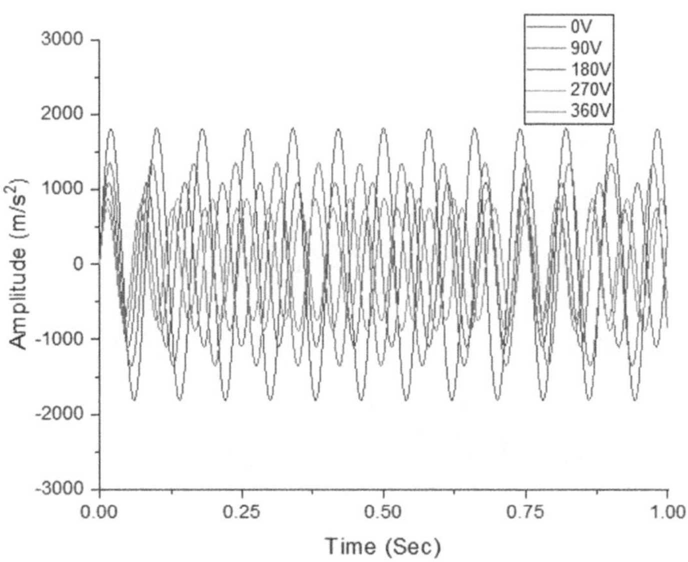

Figure 6.8 Response.

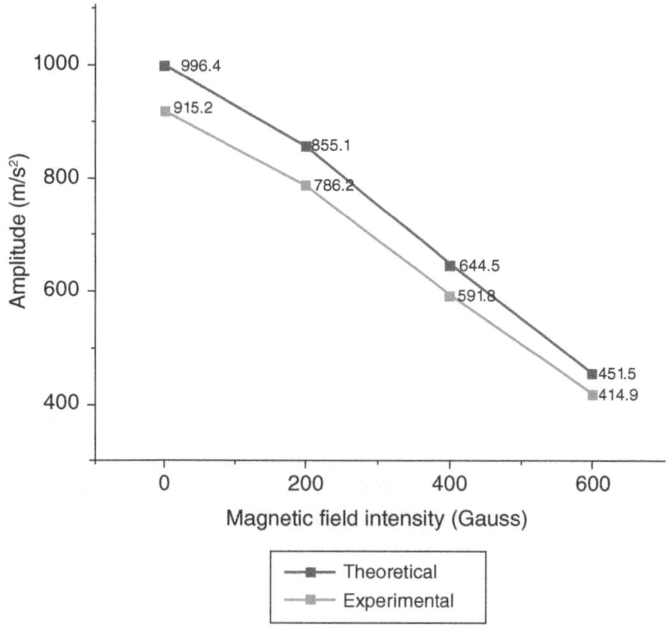

Figure 6.9 Theoretical and experimental forced vibration amplitudes for the PLA–MR fluid cantilever beam.

vibration amplitudes for the PLA–MR fluid cantilever beam, PLA–MR fluid gripper, and PLA–MFC gripper, respectively. This comparison unveils an average error percentage of 8.1%, 8.7%, and 7.7%, respectively, indicating a satisfactory alignment between theoretical predictions and experimental observations.

These findings underscore the efficacy of both MR fluid and MFC-based systems in mitigating amplitude under forced vibration conditions. The marginal discrepancy between theoretical projections and experimental results falls within an acceptable range, affirming the validity and reliability of the experimental methodology employed in this study. The significant amplitude suppression observed across all specimens highlights the promising potential of these innovative materials and configurations in vibration control applications. Further refinement and optimization efforts could enhance their efficacy and broaden their applicability in diverse engineering contexts.

6.9 CONCLUSION

This study has elucidated a notable suppression of amplitude under forced vibration conditions in PLA cantilevers and gripper fingers when subjected

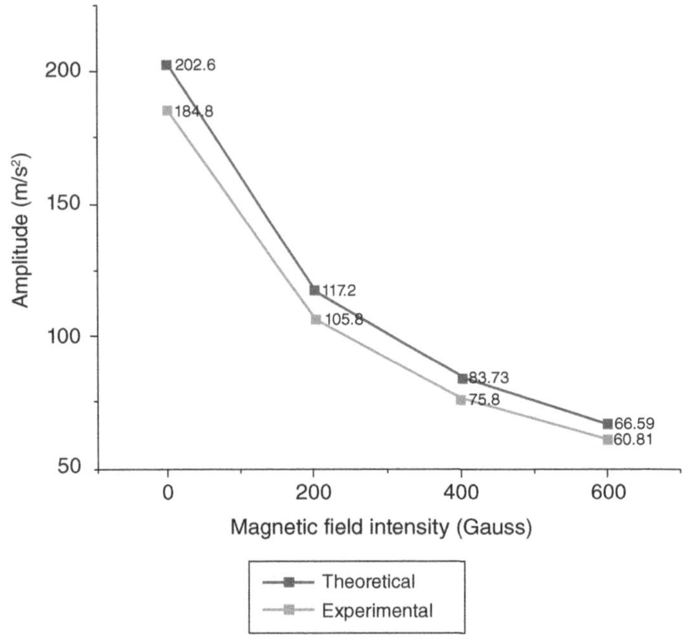

Figure 6.10 Theoretical and experimental forced vibration amplitudes for the PLA–MR fluid gripper.

to external energy stimuli. Specifically, the amplitude (acceleration) experiences a substantial reduction of 54.66% and 67.09% with the increment in magnetic field strength from 0 G to 600 G for the PLA–MR fluid cantilever beam and PLA–MR fluid gripper finger, respectively. Similarly, a significant suppression of 58.64% in amplitude (acceleration) is achieved with the elevation of the electric field from 0 V to 360 V for the PLA–MFC gripper finger.

Moreover, the comparative analysis between theoretical and experimental amplitudes reveals average error percentages of 8.1%, 8.7%, and 7.7% for the PLA MR fluid cantilever beam, PLA–MR fluid gripper finger, and PLA–MFC gripper finger, respectively. These results affirm the robustness and reliability of the experimental findings, as the observed error percentages fall within acceptable ranges. Furthermore, the consistent trend observed between theoretical predictions and experimental outcomes underscores the validity of the experimental methodology employed in this study.

In summary, this investigation underscores the efficacy of MR fluid and MFC-based systems in effectively suppressing amplitude under forced vibration conditions. The alignment between theoretical projections and

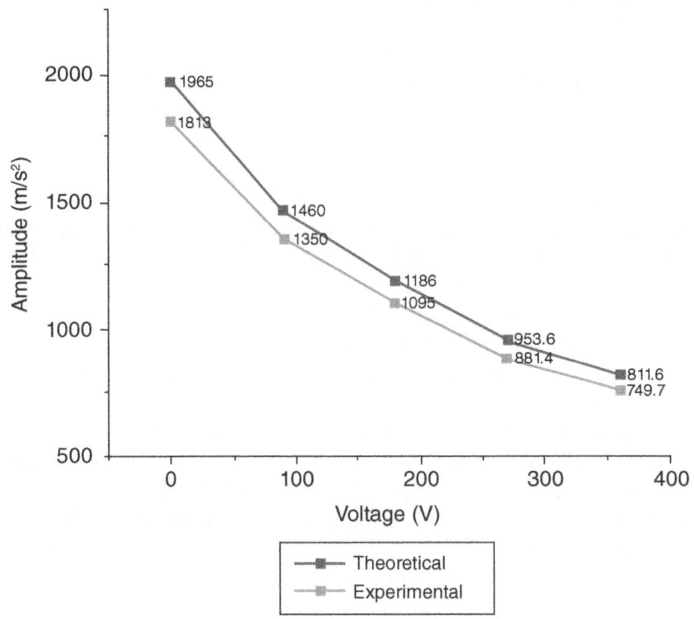

Figure 6.11 Theoretical and experimental forced vibration amplitudes for the PLA–MFC gripper.

experimental observations not only validates the experimental methodology but also emphasizes the potential of these innovative materials and configurations in advancing vibration control applications. By leveraging AI for real-time control, parameter optimization, and predictive maintenance, the study highlights the potential of smart materials in vibration control applications. Future research can further enhance these systems by integrating advanced AI algorithms for improved adaptability and performance across diverse engineering applications.

REFERENCES

1. Ramkumar G, Gnanaprakasam AJ, Thirumarimurugan M. (2021). Tunable stiffness and damping study on flexible PVC cantilever structure embedded with MR fluid. *Materials*, 14(17), 5024. https://doi.org/10.3390/ma14175024
2. Gunasekaran, R., Arul Jesu, G., Marimuthu, T. Investigation of damping characteristics on PLA robotic gripper fingers with magnetorheological fluid core and unimorph piezoelectric MFC. *Polymer Engineering and Science*, 2022, 62(7), 2102. https://doi.org/10.1002/pen.25991

3. Ramkumar, G., Gnanaprakasam, A. & Thirumarimurugan, M. (2023). Investigation of damping characteristics on PLA and PVC cantilever structures with magnetorheological fluid core. *Journal of Vibration Engineering &. Technologies*, 11, 165–173. https://doi.org/10.1007/s42417-022-00570-4
4. Ramkumar, G., Gnanaprakasam, A. J., Thirumarimurugan, M., Nandhakumar, M., Nithishmohan, M., Abinash, K., & Kishore, S. (2022). Synthesis characterization and functional analysis of magneto rheological fluid–a critical review. *Materials Today: Proceedings*, 66, 760-774.
5. Wang, C., Hong, L., & Tian, C. (2019). Preparation and properties of flame-retardant damping polyurethane and its application in floating deck. *Polymer Engineering & Science*, 59(10), 2136–2147.
6. Xu, K., Hu, Q., Wu, H., Guo, S., & Zhang, F. (2020). Achieving directional migration of hindered phenols in polymer-based damping hybrid via the construction of alternating layers. *Polymer Engineering & Science*, 60(12), 3001–3010.
7. Chang, S. H., Choi, J. K., Lee, D. G., & Kim, H. S. (1999). Damping enhancement of rubber dampers for high speed compact disk players. *Polymer Engineering & Science*, 39(9), 1642–1650.
8. Li, Y. H., Fang, B., Li, F. M., Zhang, J. Z., & Li, S. (2011). Dynamic analysis of sandwich plates with a constraining layer and a magnetorheological fluid core. *Polymers and Polymer Composites*, 19(4–5), 295–302.
9. Naji, J., Zabihollah, A., & Behzad, M. (2016). Layerwise theory in modeling of magnetorheological laminated beams and identification of magnetorheological fluid. *Mechanics Research Communications*, 77, 50–59.
10. Sripriyan, K., Ramu, M., Anantharuban, K., & Karthigha, M. (2022) Characteristic of weld bead using flat wire electrode in GMAW inline during the process: an experimental and numerical analysis. *International Journal of Pressure Vessels and Piping*, 196(January), 104623.
11. Sripriyan, K., Ramu, M., Thyla, P. R. (2023). Characterization of dissimilar welding of AISI 409 ferritic to AISI 316l austenitic stainless steel using AISI 308 filler wire under different temperature conditions. In *Advances in Additive Manufacturing and Metal Joining: Proceedings of AIMTDR 2021* (pp. 309–321). Singapore: Springer Nature Singapore.
12. Sun, S., Deng, H., Yang, J., Li, W., Du, H., Alici, G., & Nakano, M. (2015). An adaptive tuned vibration absorber based on multilayered MR elastomers. *Smart Materials and Structures*, 24(4), 045045.
13. Plattenburg, J., Dreyer, J. T., & Singh, R. (2017). Vibration control of a cylindrical shell with concurrent active piezoelectric patches and passive cardboard liner. *Mechanical Systems and Signal Processing*, 91, 422–437.
14. Sun, L., Li, W., Wu, Y., & Lan, Q. (2017). Active vibration control of a conical shell using piezoelectric ceramics. *Journal of Low Frequency Noise, Vibration and Active Control*, 36(4), 366–375.
15. Song, Z. G., Zhang, L. W., & Liew, K. M. (2016). Active vibration control of CNT-reinforced composite cylindrical shells via piezoelectric patches. *Composite Structures*, 158, 92–100.
16. Selim, B. A., Zhang, L. W., & Liew, K. M. (2017). Active vibration control of CNT-reinforced composite plates with piezoelectric layers based on

Reddy's higher-order shear deformation theory. *Composite Structures*, 163, 350–364.
17. Saviz, M. R. (2015). An optimal approach to active damping of nonlinear vibrations in composite plates using piezoelectric patches. *Smart Materials and Structures*, 24(11), 115024.
18. Oveisi, A., & Shakeri, R. (2016). Robust reliable control in vibration suppression of sandwich circular plates. *Engineering Structures*, 116, 1–11.

Chapter 7

Advancing manufacturing through data-driven smart innovation

K. Karthigadevi, G. Kalusuraman, G. Srinivasagan, and K. Sripriyan

7.1 INTRODUCTION

Manufacturers are increasingly adopting the concept of integrating the cyber and physical worlds. Various manufacturing strategies, such as Industry 4.0 in Germany, the Industrial Internet in the USA, and China's Made in China 2025 initiative, emphasize the use of modern information technologies (new IT) in manufacturing, driving the evolution of smart manufacturing [1]. Smart manufacturing focuses on converting data gathered throughout the product lifecycle into actionable insights that positively influence every aspect of the manufacturing process [2]. The volume of data generated by manufacturing systems is rapidly increasing, reaching over 1000 exabytes annually [3]. The computational analysis of this data enables better decision-making, which in turn improves the effectiveness of smart manufacturing [4]. As such, data-driven manufacturing is essential for smart manufacturing, and data has become a key factor in enhancing manufacturing competitiveness. Manufacturers are now acknowledging the strategic value of data in this context [5].

The rise of new technologies like the Internet of Things (IoT), cloud computing, mobile Internet, and artificial intelligence (AI) provides significant opportunities to support data-driven manufacturing. Innovative IoT solutions, for example, incorporate sensors into manufacturing processes to collect real-time data. Cloud computing facilitates the storage, management, and off-site analysis of this data, which can be accessed via mobile devices. AI solutions enable "smart" factories to make autonomous, timely decisions with minimal human intervention [6, 7]. Recently, there has been increased exploration of big data in manufacturing, with studies focusing on areas like industrial automation. Research has examined the role of big data in enhancing industrial competitiveness, with examples such as the impact of big data analytics on sustainable manufacturing, as highlighted by Dubey et al. [14], and the development of big data architectures for clean manufacturing and maintenance processes proposed by authors of Refs. [8, 9]. Other studies have delved into big data's applications in equipment maintenance,

 DOI: 10.1201/9781003569633-7

fault detection, fault prediction, and cost estimation [10]. Given the intelligence embedded in big data, manufacturing systems must evolve to become "smarter," enabling comprehensive monitoring, simulation, and optimization of production activities [11].

7.2 HISTORICAL PERSPECTIVES ON MANUFACTURING DATA

As shown in Figure 7.1, for a long time, information was documented on paper while manufacturing was realized by handicraft; therefore, the integration between information technology and manufacturing technology was neither beneficial nor feasible. Since the advent of ENIAC (i.e., the first electronic computer) in the 1940s, the rapid development of IT has been driving manufacturing toward informatization [12]. The first numerically controlled (NC) milling machine was developed in the 1950s, which announced that manufacturing had entered the NC era. Since the 1960s, the development of integrated circuits has paved the way for the advancement of computer hardware and software [13]. Since the 1980s, TCP/IP, local area network (LAN), World Wide Web (WWW), and search engines emerged one after another to meet the increasing needs for data storage, indexing, processing, and exchange. All these information technologies were widely applied in manufacturing [14]. As a result, many advanced manufacturing technologies were put forward, such as computer-integrated manufacturing (CIM), computer-aided design (CAD), manufacturing execution system (MES), computer-aided manufacturing (CAM), enterprise resource planning (ERP), and networked manufacturing (NM), etc. Recently, the rise of new IT (e.g., IoT, cloud computing, big data analytics, and AI) continues to revolutionize the manufacturing paradigm, leading to a series of new manufacturing concepts, for instance, manufacturing grid, cyber–physical manufacturing system, cloud manufacturing, etc. Due to the deep fusion between IT and manufacturing, the degree of manufacturing smartness is progressively elevated. As a result, the manufacturing data also becomes increasingly richer. The evolution of manufacturing data in four stages is discussed (see Figure 7.1) [15].

7.2.1 Data in the handicraft age

Prior to the First Industrial Revolution, human society had been in the manual manufacturing stage for a long time. Artifacts were predominantly designed and manufactured by artisans [16]. As the most basic form of manufacturing, handicraft activities were of low complexity. As a result, the data generated in the production process was limited, as it existed mostly in the form of human experience [17]. In addition, experience was mostly transmitted verbally from one generation to the next, primarily based on apprenticeships. The key information and data could be easily lost, making production and quality control impossible to achieve. Due to the extremely low quantity and quality, the manufacturing data generated in the handicraft age was neither emphasized nor fully exploited. However, as handcrafting involves a high level of human creativity, even today, it is used to manufacture luxury products (e.g., jewelry, watches, leather bags) [18].

7.2.2 Data in the machine age

The machine age consisted of two phases. As a result of the first industrial revolution, machines were employed as production tools in the early factories, leading to a significant increase in the scale of manufacturing. During this period, the relationship between humans and machines in production was highly complementary (i.e., early machines could only be operated by skilled operators to deliver their functions) [19]. Therefore, manufacturers began to emphasize two kinds of manufacturing data: worker-related data and machine-related data. As a result of the Second Industrial Revolution (or the Technological Revolution), machine tools and interchangeable parts were widely incorporated into the "new" manufacturing process (e.g., the Bessemer process) in modern factories, leading to significant increases in manufacturing efficiency, and the manufacturing paradigm shifted to the mass production model [20, 21]. The Second Industrial Revolution triggered some notable changes in the way data was processed. Because of the division of work between managers and workers, manufacturing data was increasingly handled by educated managers. Moreover, managers began to employ more systematic methods to document and analyze manufacturing data. The raw data was extensively recorded in written documents (e.g., instructions, logbooks, notes, and charts) rather than stored in human memory. Scientific methods were used to determine the dependency relationships between different datasets. During this period, manufacturers began to exploit manufacturing data for cost reduction, quality control, and inventory management [22]. In particular, statistical models were introduced to analyze a variety of quality-related manufacturing data, such as production planning, throughput yield, product quality, failure rate, raw material consumption, and scrap rate. In summary, in the machine age, although a larger quantity of manufacturing data was analyzed through scientific methods, data was still handled manually by human operators (i.e., managers), as opposed to computers. Therefore, the utilization rate of manufacturing data remained relatively low [23].

7.2.3 Data in the information age

In the information age (or the digital age), information technologies were widely applied in manufacturing processes. Consequently, the quantity of manufacturing data that companies were able to collect grew exponentially [24]. A few factors contributed to this growth in data. First, information systems (e.g., CRM, MES, ERP, SCM, PDM) were widely employed by manufacturers to facilitate production management. Second, computer systems (such as CAD, CAE, CAM, and FEA) were widely used to aid the creation, simulation, modification, and optimization of new products as well as manufacturing processes. Third, industrial robots and automatic machinery were commonly used in modern factories [25, 26]. Electronic devices and digital computers were increasingly being used to automatically control production equipment. The evolvements in information technologies paved the way for manufacturers to achieve meeting customer needs better, quicker, and cheaper. In the information age, data was stored in computer systems and managed by information systems [27]. For example, customer data (e.g., home address, phone number, demographics), sales data (e.g., type, quantity, price, and shipping date of finished products), supply chain data (e.g., type, quantity, price, and supplier of raw materials), financial data (e.g., assets, real property, tangible property, utility, intangible property), production planning data, bill of materials, inventory data (e.g., type, quantity, location of material and finished products in the warehouse), and maintenance data are all managed by CRM, MES, ERP, SCM, PLM, etc. [28]. Therefore, it could be easily exchanged among different departments or organizations.

Figure 7.1 presents a timeline highlighting technological advancements in both information technology and manufacturing technology from 1800 to 2010 [29, 30]. On the information technology side, key milestones include the telegraph and telephone (around 1800), ENIAC and the assembly line (around 1950), ARPANET (1950–1970), the introduction of TCP/IP and local area networks (1980), the WWW and search engines (1990), and the rise of IoT and cloud computing (2000). Meanwhile, manufacturing technology features the advent of handicrafts and the steam engine (around 1800), numerical control (1950), automation and CIM (1970), CAD (1980), CAM and ERP (1990–2000), and the emergence of cyber–physical manufacturing systems and Industry 4.0 (2010) [31]. The timeline effectively illustrates the parallel evolution of both fields, showcasing significant milestones in their development. Abbreviations like MES and ERP are also defined.

7.2.4 Data in the big data age

Along with the rise of IoT technologies, cloud computing, big data analytics, AI, and other technological advances came the age of big data. In manufacturing, big data refers to large amounts of multi-source, heterogeneous data generated throughout the product lifecycle, characterized by 5 Vs: high volume (i.e., huge quantities of data), variety (i.e., the data itself comes

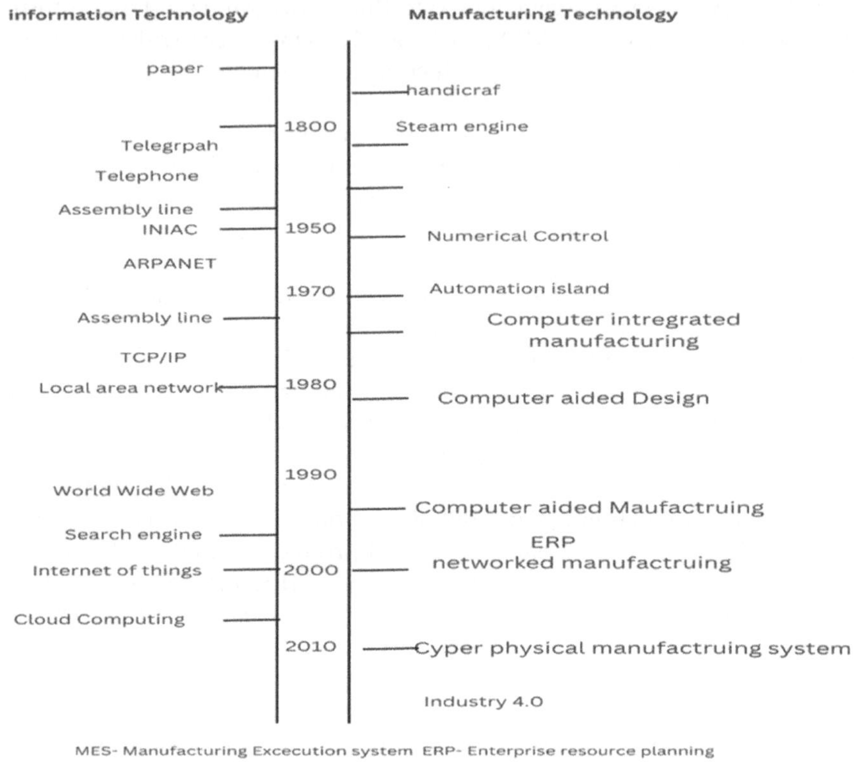

Figure 7.1 Evaluation of data in manufacturing.

in different forms and is generated by diverse sources), velocity (i.e., the data is generated and renewed at very high speed), veracity (i.e., the data is associated with a level of bias, inconsistency, incompleteness, ambiguities, latency, noises, and approximation), and value (i.e., huge value hidden in the data) [32]. Big data generated by manufacturing processes can be classified according to the following categories:

1. Management data collected from manufacturing information systems (e.g., MES, ERP, CRM, SCM, and PDM). Information systems possess a variety of data related to product planning, order dispatch, material management, production planning, maintenance, inventory management, sales and marketing, distribution, customer service, and financial management.
2. Equipment collected data from smart factories by Industrial IoT technologies, which includes data related to real-time performance,

operating conditions, and the maintenance history of production equipment.

3. User data collected from Internet sources such as e-commerce platforms (e.g., Amazon, Walmart, and Taobao) and social networking platforms (e.g., Twitter, Facebook, LinkedIn, and YouTube). This type of data encompasses user demographics, user profiles, and user preferences toward products/services, as well as user behavior (e.g., data about online browsing, searching, purchasing, and reviewing history).
4. Product data collected from smart products and product-service systems by IoT technologies, including product performance, context of use (e.g., time, location, and weather), environmental data (e.g., temperature, humidity, and air quality), and user biological data.
5. Public data collected from governments through open databases. Such datasets deal with data related to intellectual property, civic infrastructure, scientific development, environmental protection, and healthcare. For manufacturers, public data can be used to guarantee that manufacturing processes and manufactured products strictly comply with government regulations and industry standards.

In the big data age, empowered by the new ITs, manufacturers' ability to collect, store, and process data is significantly enhanced. Recently, there have emerged several cost-effective and flexible data collection, storage, and processing solutions such as the IoT and cloud computing [33]. As a result, manufacturing enterprises of different scales, even including SMEs, can benefit from the value of data. In manufacturing, effective analysis of big data enables manufacturers to deepen their understanding of customers, competitors, products, equipment, processes, services, employees, suppliers, and regulators.

7.3 LIFECYCLE OF MANUFACTURING DATA

Data is a key enabler for smart manufacturing. However, data is not useful unless it is "translated" into concrete information content and context that can be directly understood by users. Generally, before getting the concrete information from data, the data need to pass through multiple steps. The complete journey of data collection, transmission, storage, pre-processing, filtering, analysis, mining, visualization, and application can be referred to as the "data lifecycle." Manufacturing data is exploited at various points in the data lifecycle [34]. As illustrated in Figure 7.2, a typical manufacturing data lifecycle consists of data collection, transmission, storage, processing, visualization, and application.

7.3.1 Data sources

The volume of data collected across the entire manufacturing value chain and product lifecycle is increasing at an unprecedented rate. As discussed in Section 2.4, the manufacturing data comes from equipment, products, human operators, information systems, and networks [35].

7.3.2 Data collection

Data from different sources is collected in a variety of ways. Above all, it is collected by means of the IoT, whereby equipment and product data can be instantly collected through smart sensors, radio frequency identification (RFID), and other sensing devices, making it possible to monitor equipment and product health in real time [36]. For instance, built-in sensors make it possible to continuously measure, monitor, and report the ongoing operational status of manufacturing equipment and products, such as temperature, pressure, and vibration. RFID enables the automatic identification, tracking, and management of many workpieces, as well as the materials necessary for production. Moreover, the emerging mobile Internet paves the way for user data collection through smart terminals (e.g., devices like PCs, phones, laptops, and tablets). Through software development kits (SDKs) or application programming interfaces (APIs), for example, basic user data can be collected, including the number of users, user profiles, location, and time [37]. In addition, web crawling is a widely used data acquisition technique for collecting public data based on certain conditions predefined by engineers and AI. Web crawling refers to the technology of deploying "crawlers" (i.e., computer programs) to browse public web pages and collect desirable information [38]. The web crawling technology enables manufacturers to acquire public data in an automatic and efficient manner. Finally, management data from manufacturing information systems can be acquired and used at any time through database technologies.

7.3.3 Data storage

The large volume of data collected from manufacturing processes must be securely stored and effectively integrated. The various types of manufacturing data can be classified into structures (e.g., digits, symbols, tables), semi-structured (e.g., trees, graphs, XML documents), and unstructured data (e.g., logs, audios, videos, images). Traditionally, manufacturing enterprises focused heavily on structured data storage, as it was difficult to directly manage unstructured data within enterprise databases. Object-based storage architecture enables collections of data to be stored and managed as objects; this provides a more flexible solution for integrating semi-structured and unstructured data [39]. Also, through cloud computing, data storage can be achieved in a highly cost-effective, energy-efficient, and flexible fashion. In addition, by virtue of cloud services, the distribution and heterogeneity of data are shielded, enabling a highly scalable and shareable mode of data storage.

7.3.4 Data processing

Data processing refers to a series of operations conducted to discover knowledge from a large volume of data. Data must be converted to information and knowledge for manufacturers to make informed and rational decisions. Above all, data must be carefully preprocessed to remove redundant, misleading, duplicate, and inconsistent information. Specifically, data cleaning involves the following activities: missing value, format, duplicate, and garbage data cleaning. Data reduction is the process of transforming the massive volume of data into ordered, meaningful, and simplified forms by means of feature or case selection. After data reduction has been completed, the cleaned and simplified data is exploited through data analysis and mining to generate new information [40]. The effectiveness of data analysis can be significantly enhanced through a variety of techniques, including machine learning, large-scale computing, and the use of forecasting models. Some advanced data mining methods include clustering, classification, association rules, regression, prediction, and deviation analysis. Through the above data processing efforts, understandable knowledge can be derived from a large quantity of dynamic and ambiguous raw data.

7.3.5 Data visualization

Visualization is intended to clearly convey and communicate information through graphical means, enabling end users to understand data in a much more explicit fashion. The most used visualization techniques include statements, charts, diagrams, graphs, and virtual reality [41]. Real-time data can be visualized online via users smart terminals. Through visualization, the results of data processing are made more accessible, straightforward, and user-friendly [42].

7.3.6 Data transmission

Data is continuously flowing among different information systems, cyber–physical systems, and human operators. Data transmission, therefore, plays a critical role in maintaining communications and interactions among distributed manufacturing systems and resources [43]. The recent advances in IoT, Internet, and communication networks substantially consolidated the technological foundation of real-time, reliable, and secure transmission of different types of data. As a result, distributed manufacturing resources can be effectively integrated almost anytime and anywhere.

7.3.7 Data applications

Data has entered almost all aspects of daily production and operation in manufacturing enterprises. First, during the design phase, through data analytics,

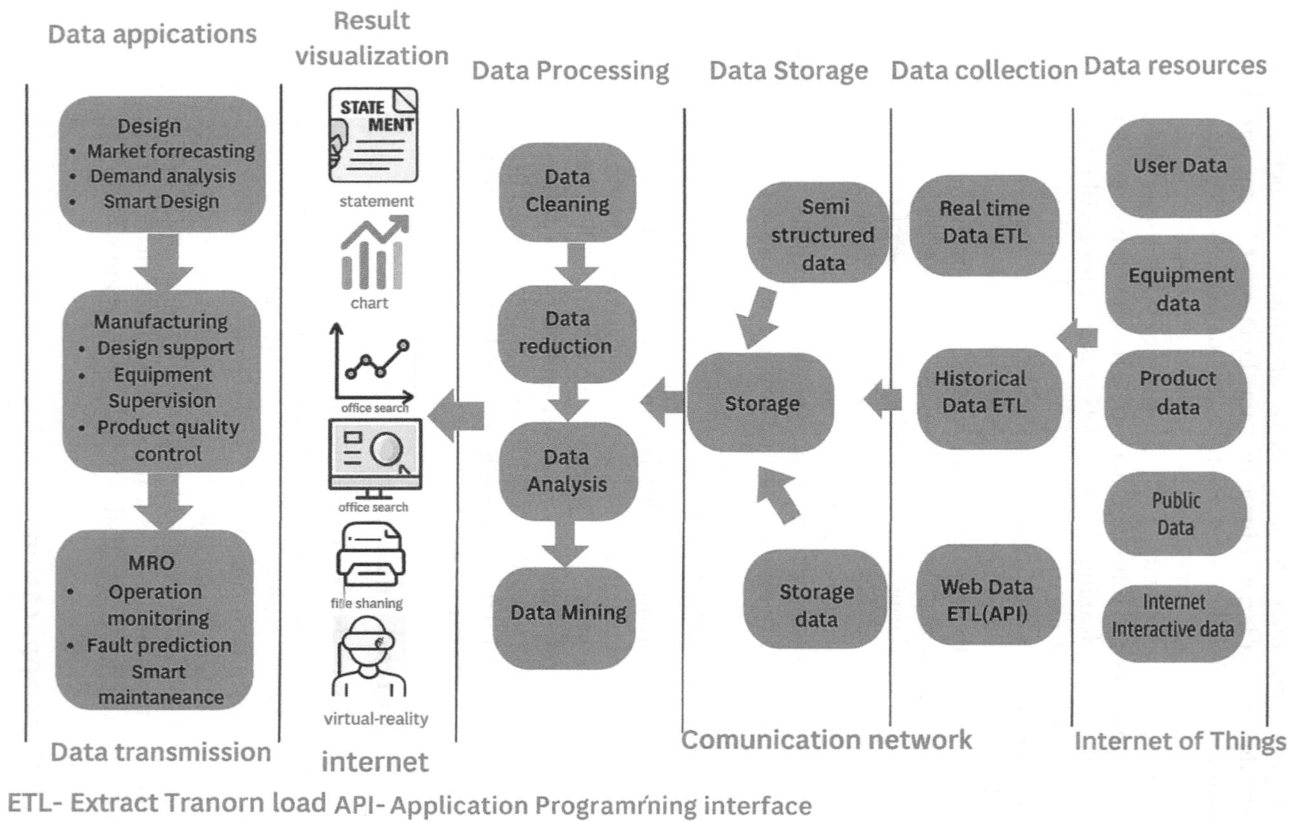

Figure 7.2 Data life cycle.

new insights are revealed about customers, competitors, and markets. Based on the understanding developed through data analytics, designers can accurately and rapidly translate customer voices to product features and quality requirements [44]. As a result, manufacturers will become closer to customers and more agile in terms of coping with a dynamic, changing market.

Figure 7.2 represents the manufacturing data life cycle, which involves the stages through which data moves within a manufacturing system, starting from its creation to its eventual disposal. It begins when data is generated during processes such as design, production, and operations, capturing crucial information like machine performance, materials used, and product details. This data is then stored, processed, and analyzed to enhance decision-making, boost efficiency, and optimize production [45]. As it progresses, the data is integrated across various systems like MES and ERP to ensure seamless information flow. Eventually, data is archived for future use or regulatory needs, and outdated data is removed to maintain system performance [46]. The life cycle ensures data integrity, accessibility, and security throughout the manufacturing process.

7.4 DATA-DRIVEN SMART MANUFACTURING

7.4.1 The connotations of data-driven smart manufacturing

Manufacturing enterprises utilize big data analytics to exploit the data from manufacturing to refine the manufacturing process, improving the flexibility and smart level of manufacturing. By taking full advantage of manufacturing data, manufacturing shifts from primary processes to smart processes, thus improving production efficiency and product performance [47].

7.4.1.1 Data-driven smart manufacturing framework

The manufacturing data is collected, stored, processed, and analyzed by means of big data technologies. As a result, the degree of manufacturing intelligence can be significantly elevated.

The data-driven smart manufacturing framework consists of four modules, namely:

Manufacturing module: This module accommodates different kinds of manufacturing activities. It consists of a variety of information systems and manufacturing resources, which can be summarized as man–machine–material–environment [48]. The inputs to this module are raw materials, whereas the outputs are finished products. During the input–output transformation process, various data is collected from human operators, production equipment, information systems, and industrial networks.

Data driver module: This module provides the driving force for smart manufacturing throughout the different stages of the manufacturing data lifecycle. As inputs, the data from the manufacturing module is transmitted to cloud-based data centers for further analysis [49]. Explicit information and actionable recommendations derived from the raw data are used to direct actions (e.g., product design, production planning, and manufacturing execution) in the manufacturing module.

Real-time monitoring module: This module monitors the manufacturing process in real time to ensure product quality. Driven by the data driver module, it analyzes the real-time running status of manufacturing facilities. Manufacturers can stay updated on changes in the manufacturing process, enabling the development of optimal operational control strategies [50].

Problem processing module: This module identifies and predicts emerging problems (e.g., equipment faults or quality defects), diagnoses root causes, recommends solutions, and estimates effectiveness. It also evaluates potential impacts on other manufacturing activities. Based on real-time information and analysis of historical and ongoing data, either human operators or AI applications can make informed decisions to address current problems and prevent similar issues in the future.

The structured process of data collection, integration, storage, analysis, visualization, and application is generally applicable for various industries. This data-driven smart manufacturing framework is universally valuable. Depending on resource availability, both SMEs and large companies can adopt different strategies to achieve data-driven smart manufacturing at different scales [51]. For example, unlike larger companies that can afford exclusive cloud infrastructure, SMEs can use third-party cloud computing services such as those provided by Amazon and Alibaba. Regardless of how and where data is processed, the key value propositions of data-driven manufacturing are the same for both SMEs and large companies. Manufacturing data helps decision-makers understand changes quickly, make accurate judgments, and develop rapid response measures to address issues. This leads to closely coordinated production plans, manufacturing activities, and resources to promote smart manufacturing.

7.4.1.2 Characteristics of data-driven smart manufacturing

The data-driven smart manufacturing shares the following five characteristics:

Customer-centric product development: It uses user data to customize product design. User demographics, demands, preferences, and behaviors can be precisely quantified by using big data analytics to create more personalized products and services.

Self-organization: It uses manufacturing resources and task data for smart production planning. Production plans are created based on internal and external data from different manufacturing sites, with resources chosen to form the optimal configuration to meet all demands.

Self-execution: It uses data from the manufacturing process for precise control. Raw materials and parts can be automatically sent to manufacturing sites, and equipment can automatically machine or assemble materials when needed.

Self-regulation: It uses real-time status data for manufacturing process monitoring. The system can automatically respond to unexpected events, such as resource shortages or changes in manufacturing tasks, making its behaviors controllable through both human operators and AI systems.

Self-learning and self-adaptation: It uses historical and real-time data for proactive maintenance and quality control. Machine faults and quality defects can be predicted and prevented before they occur, allowing the system to adapt proactively to potential issues.

In summary, data-driven smart manufacturing offers a wide range of services to manufacturing enterprises, improving both manufacturing efficiency and product performance (Figures 7.3 and 7.4).

7.4.2 Data-driven smart manufacturing applications

Manufacturing transforms raw material inputs into finished products and value-added services through the coordination of manufacturing facilities, resources, and activities. Some of the most promising applications that can be implemented during the manufacturing process include smart design, smart planning, materials distribution and tracking, manufacturing process monitoring, quality control, and smart equipment maintenance.

7.4.2.1 *Smart design*

Design plays a crucial role in determining most of a product's manufacturing costs. In the era of big data, product design is evolving toward a more data-driven approach. Product design begins by gathering and understanding customer demands, behaviors, and preferences. Data can be collected from Internet and IoT sources. Internet data is becoming increasingly available as customers share their experiences on social networking sites, e-commerce platforms, and product review sites [52]. IoT sources also provide rich user data (e.g., biological data, behavioral data, and user–product interactions)

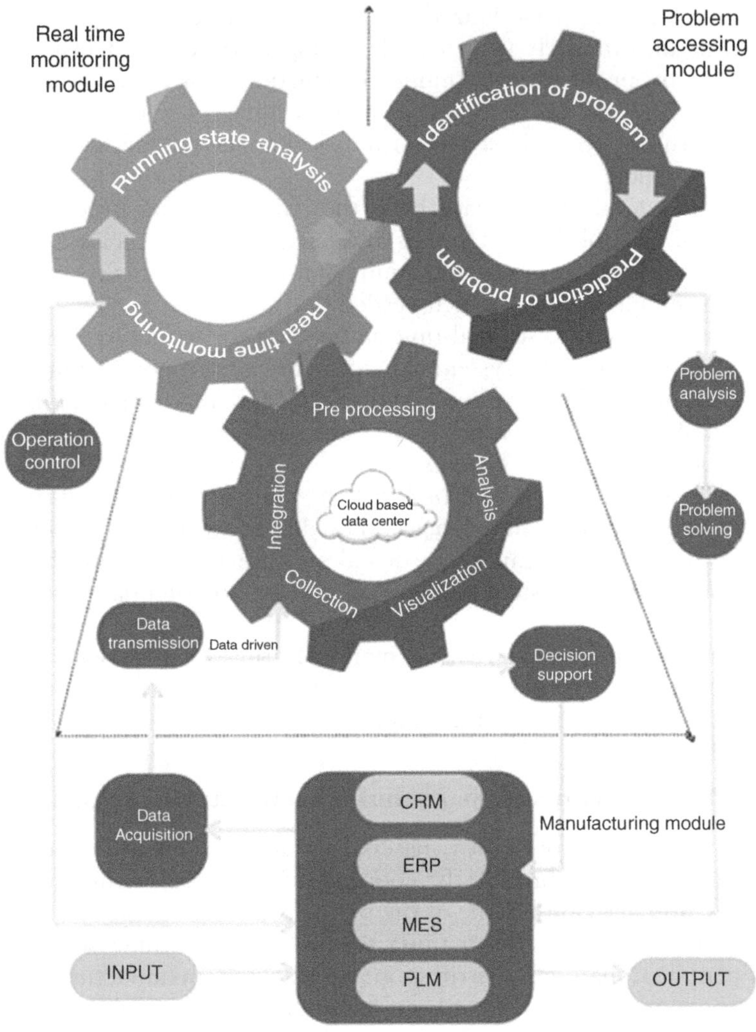

Figure 7.3 Framework of data-driven smart manufacturing.

from smart devices like smartphones and wearable devices. This data enables manufacturers to translate customer feedback into product features and quality requirements. Moreover, the use of cloud-based high-performance computing and big data analytics helps accelerate computational tasks, such as analyzing market preferences and customer demands, while reducing costs compared to traditional methods like surveys and interviews.

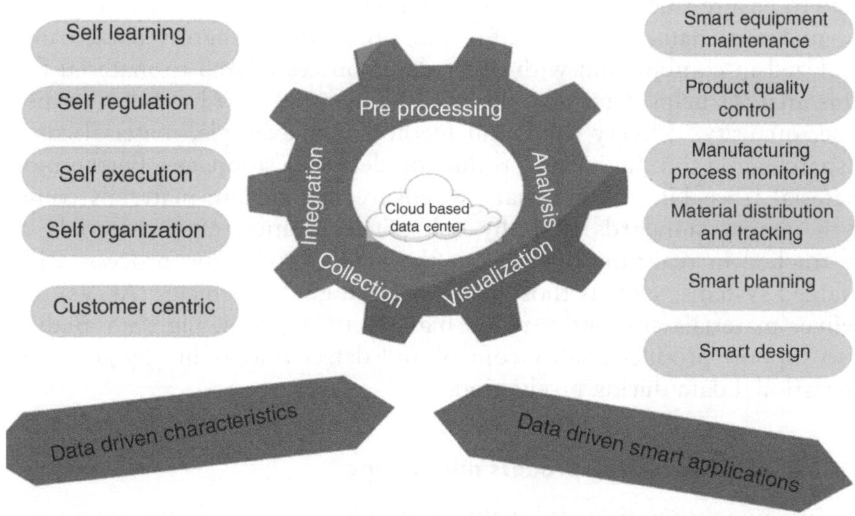

Figure 7.4 Applications and characteristics of data-driven smart manufacturing.

7.4.2.2 Smart planning and process optimization

Before manufacturing a product, effective production planning is essential to determine the capacity of manufacturing facilities and the availability of resources and materials. Big data analytics can make production planning and scheduling smarter. Various data, such as customer orders, resource status, production capacities, supply chain data, and inventory information, are analyzed using big data techniques [53]. With this information, manufacturing resource supply-and-demand matching and scheduling can be quickly optimized. Additionally, intelligent optimization algorithms are used to develop production plans and determine the best configuration of manufacturing resources and the necessary execution procedures. Process optimization is another critical element before production begins [54]. Big data analytics helps assess and optimize technological processes by analyzing historical data and data related to specific processing steps. By understanding the correlation between technological parameters and their effects on yield and quality, production processes can be adjusted for higher productivity and better product quality while reducing costs.

7.4.2.3 Material distribution and tracking

Material distribution is determined through production planning and ongoing production progress, as well as on-site urgency requirements.

Ideally, materials should be delivered to the right equipment at the right time to ensure smooth operations. A variety of material-related data, such as inventory data, logistics data, and production progress data, can be analyzed in combination with other data sources related to material flow. This analysis helps determine the optimal material distribution in terms of type, quantity, delivery time, and method. For example, material can be dispatched on time, avoiding production delays or stoppages. Furthermore, material traceability is necessary to ensure that certain materials comply with quality standards [55]. By using identification tags, materials can be tracked in real-time throughout the entire production process. RFID-enabled systems, such as those in automated guided vehicles (AGVs), help deliver materials efficiently within manufacturing sites. Big data analytics also support product quality control and defect traceability by generating operational data during production.

7.4.2.4 Manufacturing process monitoring

The manufacturing process involves multiple factors, including equipment, materials, environmental conditions, and technological parameters, which can all influence product quality. These factors can interact with each other, making it essential to monitor different stages of the manufacturing process in real time. Big data provides effective support for manufacturing process monitoring [56]. By analyzing data patterns and using predictive analytics, big data helps determine the acceptable ranges for each factor. If a factor deviates from its acceptable range, the system flags the issue, sending alerts and recommendations to operators for timely adjustments. This helps maintain consistency in the manufacturing process. For example, production abnormalities, such as order delays, are often caused by events like equipment failures, material shortages, or operational deviations. These events typically follow patterns that can be detected by time-series data, including material consumption, energy usage, and machine performance metrics (e.g., rotation rate, vibration, torque). As such data is time-dependent, it is challenging to process using static models or traditional data analysis methods [57]. However, big data analytics, which can handle massive and multi-source data efficiently, can be used to build early-warning models for production abnormalities. By analyzing patterns and trends in the data, big data can predict when production issues may occur, allowing for timely adjustments to the manufacturing process (Figure 7.5).

7.4.2.5 Product quality control

Various data-driven techniques are being developed to enhance quality control in smart manufacturing. Sensors, RFIDs, and machine vision applications are commonly used to gather product quality data, such as

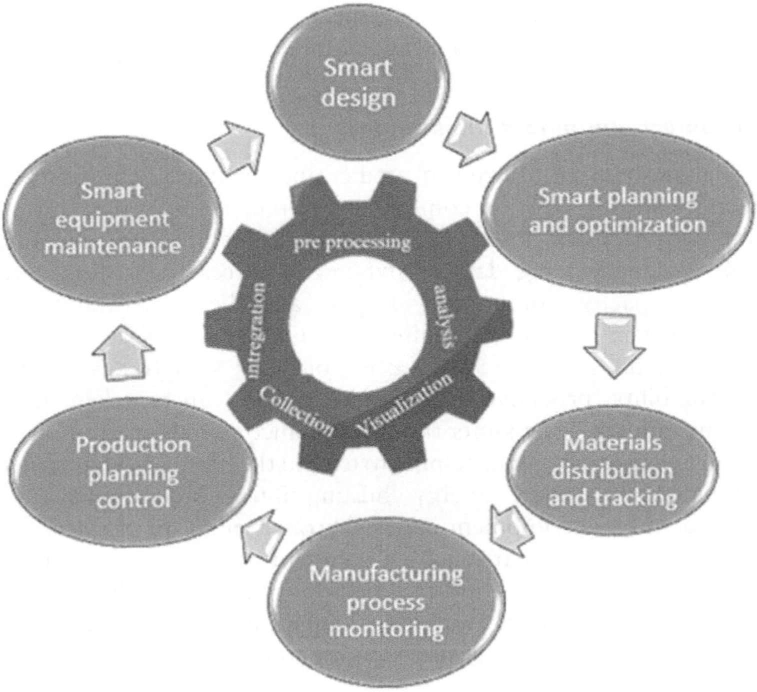

Figure 7.5 Application of data-driven smart manufacturing.

geometric parameters (e.g., thickness, length, and surface roughness), location parameters (e.g., coordinates), tolerance parameters (e.g., concentricity), and machining parameters (e.g., pressure, speed, temperature, and machining time). Big data analytics can facilitate comprehensive quality monitoring, early detection of quality defects, and rapid identification of root causes. By analyzing historical data and process condition data collected from machines and their operating environments, it is possible to predict the relationship between certain conditions and quality defects. For instance, binary classification methods can be employed to categorize quality conditions and determine their link to product failures. Bayesian inference techniques can also be applied to identify key parameters influencing product defects, along with their optimal ranges. Additionally, root cause analysis, combined with weighted association rule mining, helps in identifying the causes of product failures. This approach not only detects and removes defective products but also helps eliminate or control factors that contribute to quality issues [58]. Furthermore, by incorporating machine learning, big data analytics enable manufacturing enterprises to develop case-based reasoning capabilities, allowing insights from one quality control case to be applied to future scenarios, preventing the recurrence of similar problems. As a result, quality management is integrated into every stage of the manufacturing process, from raw materials to the final product.

7.4.2.6 Smart equipment maintenance

Data analytics can be utilized to predict and diagnose equipment failures and component lifespan. By combining equipment status data from smart sensors with domain knowledge, historical maintenance records, and previous experience, big data analytics can forecast the degradation of equipment capacity, component lifespan, and the cause and severity of faults. Moreover, patterns of equipment faults whether seasonal, periodic, or combinational can be detected using big data analytics. With this predictive capability, proactive maintenance actions can be taken to prevent equipment failures. This shifts the maintenance paradigm from a reactive approach to a proactive one, helping to extend the lifespan of equipment and reduce maintenance costs. Energy consumption data is another important factor in identifying equipment issues. Through multi-dimensional analysis of energy consumption patterns, big data can uncover energy fluctuations and anomalies, which may indicate potential faults. By adjusting production processes, equipment, and energy supply in real time based on these insights, manufacturers can ensure smooth and optimized operations.

7.4.2.6.1 Wafer production line

Silicon wafers are essential components in crystalline silicon photovoltaic cells, which play a crucial role in enhancing solar energy technology. The wafer production process starts with the input of silicon ingots and ends with the output of silicon wafers. The process includes several key stages, such as loading material, cutting, chamfering and polishing, viscos, slicing, degumming and cleaning, sorting, and packaging. The production line consists of various pieces of equipment associated with each of these stages.

Throughout the production process, diverse types of data from multiple sources are continuously collected. This data is integrated with order and production plan information [59]. By leveraging big data analytics, intelligent algorithms, and predictive models, the data is analyzed to optimize the manufacturing process. As a result, big data analytics help enable efficient material allocation, real-time tracking, predictive maintenance, and energy efficiency management.

For material distribution and tracking, RFID tags are attached to materials, and external readers are used to gather material data. The data is recorded as a 9-tuple, such as:

Material = {ID, time, location, batch, type, quantity, sender, receiver, item code}.

Every time the material passes through an RFID reader, this data is collected. A significant volume of data is generated from the silicon ingot input to the silicon wafer output. Material identification is achieved through data fusion technology, which ensures that recognition confidence meets a predefined threshold. If the result falls below this threshold, an alarm is triggered on mobile devices (e.g., smartphones and tablets) for manual intervention. Additionally, operators can monitor the material's status in real time, including its processing location and stage, at any given moment. A dynamic material distribution scheme is implemented to track the location, batch, type, and quantity of material to be delivered. Delivery instructions and routes are then visualized for operators on their mobile devices. Furthermore, the entire material dataset covering when, where, and through which production processes the material passed can be reviewed at any time.

For fault diagnosis and prediction, production equipment is fitted with sensors to monitor various parameters, such as location, weight, temperature, humidity, vibration, and flow rate. Real-time data is analyzed to identify which equipment requires service, repairs, or even replacement [60]. For instance, vibration data from multiple sensors is used to assess the operational state of a multi-wire slicing machine. Initially, noise and redundant data are filtered using a wavelet transform module maxima denoising technique. Next, a feature extraction method based on attribute reduction is employed to derive key parameters from the vibration data.

7.4.2.6.2 Smart failure diagnosis and energy efficiency management

Based on the backpropagation (BP) neural network, a smart failure diagnosis system is implemented. Under normal conditions, the vibration signal of the equipment exhibits a stable pattern. However, when the equipment experiences wear or unexpected faults, the vibration signal deviates from this normal pattern, automatically triggering a warning sent to the operator via mobile terminals. By analyzing the vibration signal, equipment anomalies can be predicted and diagnosed in real time.

For energy efficiency management, energy consumption during the manufacturing process is measured using smart meters installed on each production machine. The collected data undergoes multi-dimensional analysis to optimize energy use. Initially, hierarchical cluster analysis is applied to identify energy consumption patterns over specific periods, enhancing energy efficiency [61]. The advantage of this method is that it does not require extensive modeling or computational efforts. By examining energy consumption over daily, weekly, monthly, and yearly intervals, trends in energy use can be identified.

Further predictive analytics of energy consumption are conducted using the autoregressive integrated moving average (ARIMA) algorithm. The representation of the energy consumption records over different time spans (monthly, weekly, and daily). This allows manufacturers to observe short-term and long-term energy consumption trends, facilitating better energy planning. Additionally, comparing real-time data with historical records enables the evaluation of overall consumption patterns and trends, which is crucial for determining when to carry out comprehensive overhauls. Unusual fluctuations in energy consumption can also signal potential abnormalities in the production process.

7.5 CONCLUSION AND FUTURE WORK

The volume of dynamically changing data generated throughout the product lifecycle is steadily growing. This data, if properly utilized, can significantly enhance manufacturing efficiency. This paper has contributed to the field of smart manufacturing in three key areas: (1) historical perspective, reflecting the evolution of manufacturing data across four distinct eras: the handicraft age, the machine age, the information age, and the big data age; (2) development perspective, outlining the lifecycle of big manufacturing data, which includes phases such as generation, collection, transmission, storage, integration, processing, analysis, visualization, and application; and (3) future perspectives, exploring the role of data analytics in manufacturing, with a focus on promising smart manufacturing applications. However, several limitations need to be addressed. First, current data collection technologies are not fully optimized for smart data perception, particularly when managing heterogeneous devices with different communication protocols and interfaces. Second, while cloud-based storage and analytics have proven feasible, issues such as network unreliability, bandwidth constraints, and unacceptable latency times limit their use in real-time, low-latency applications. Third, while integrating the physical and cyber worlds is a hallmark of smart manufacturing, most research has concentrated on data from the physical world, overlooking virtual model data.

Looking ahead, there are several promising avenues for future research:

Data perception and collection technologies: Incorporating key technologies, such as IoT gateways or industrial Internet hubs, into the data-driven smart manufacturing framework could enhance compatibility with heterogeneous devices, improving data collection and transmission.

New storage and processing technologies: Integrating fog computing and edge computing into the framework can extend computing, storage, and networking capabilities beyond the cloud, reducing bandwidth requirements, latency, and service downtime.

Digital twin technologies: Incorporating digital twin technologies into the framework can enable real-time, two-way, coevolving mappings between physical objects and their digital representations. This enhances cyber–physical integration, making data-driven smart manufacturing more responsive, adaptable, and predictive.

REFERENCES

1. Tao F, Qi Q. New IT driven service-oriented smart manufacturing: framework and characteristics. *IEEE Trans Syst Man Cybern Syst* 2017;47(7):1771–83, http://dx.doi.org/10.1109/TSMC.20172723764
2. O'Donovan P, Leahy K, Bruton K, O'Sullivan DTJ. A big industrial data pipeline for data-driven analytics maintenance applications in large-scale smart manufacturing facilities. *J Big Data* 2015;2(25):1–26, http://dx.doi.org/10.1186/s40537-015-0034-z
3. Yin S, Kaynak O. Big data for modern industry: challenges and trends. *Proc IEEE* 2015;103(2):143–6.
4. Shao G, Shin SJ, Jain S. Data analytics using simulation for smart manufacturing. In: *Proceedings of the 2014 Winter Simulation Conference* (pp. 2192–203), 2014. IEEE.
5. Kusiak A. Smart manufacturing must embrace big data. *Nature* 2017;544(7648):23–5.
6. Mourtzis D, Vlachou E, Milas N. Industrial big data as a result of IoT adoption in manufacturing. *Procedia CIRP* 2016; 55:290–5.
7. Tao F, Zuo Y, Xu LD, Zhang L. IoT-based intelligent perception and access of manufacturing resources toward cloud manufacturing. *IEEE Trans Ind Inf* 2014;10(2):1547–57.
8. Hashem IAT, Yaqoob I, Anuar NB, Mokhtar S, Gani A, Khan SU. The rise of big data on cloud computing: review and open research issues. *Inf Syst* 2015;47:98–115.
9. Tao F, Cheng Y, Xu LD, Zhang L, Li BH. CCIoT-CMfg: cloud computing and internet of things-based cloud manufacturing service system. *IEEE Trans Ind Inf* 2014;10(2):1435–42.
10. Lee MD, Butavicius MA, Reilly RE. Visualizations of binary data: a comparative evaluation. *Int J Hum Comput Stud* 2003;59(5):569–602.
11. Wuest T, Weimer D, Irgens C, Thoben KD. Machine learning in manufacturing: advantages, challenges, and applications. *Prod Manuf Res* 2016;4(1):23–45.
12. Obitko M, Jirkovsky V, Bezdicek J. Big data challenges in industrial automation. In V. Mařík, J. L. Martinez Lastra, & P. Skobelev (Eds.), *Industrial Applications of Holonic and Multi-Agent Systems: 6th International Conference, HoloMAS 2013, Prague, Czech Republic, August 26-28, 2013* (Lecture Notes in Computer Science, Vol. 8062, pp. 305–16). Springer.
13. Galletti A, Papadimitriou DC. How big data analytics are perceived as a driver for competitive advantage: a qualitative study on food retailers; 2013, pp. 1–58. Master Thesis.
14. Dubey R, Gunasekaran A, Childe SJ, Wamba SF, Papadopoulos T. The impact of big data on world-class sustainable manufacturing. *Int J Adv Manuf Technol* 2016;84(1–4):631–45.

15. Zhang Y, Ren S, Liu Y, Si S. A big data analytics architecture for cleaner manufacturing and maintenance processes of complex products. *J Clean Prod* 2017;142:626–41.
16. Bahga A, Madisetti VK. Analyzing massive machine maintenance data in a computing cloud. *IEEE Trans Parallel Distrib Syst* 2012;23(10):1831–43.
17. Lee H. Framework and development of fault detection classification using IoT device and cloud environment. *J Manuf Syst* 2017;43:257–70.
18. Munirathinam S, Ramadoss B. Big data predictive analytics for proactive semiconductor equipment maintenance. In: *Proceedings of the 2014 IEEE International Conference on Big Data (Big Data), Washington, DC, USA, October 27–30* (pp. 893–902), 2014. IEEE.
19. Chan SL, Lu Y, Wang Y. Data-driven cost estimation for additive manufacturing in cyber manufacturing. *J Manuf Syst* 2018; 46:115–26.
20. Schleich B, Anwer N, Mathieu L, Wartzack S. Shaping the digital twin for design and production engineering. *CIRP Ann Manuf Technol* 2017;66(2017):141–4.
21. Karuthapandi S, Murugan R, Palani PK. Study and analysis of the macrostructure characteristics in FCAW with the use of a flat wire electrode and by optimizing the process parameter using the Taguchi method and regression analysis. *High Temp Mater Processes* 2016;20(3):197–224.
22. Hu SJ. Evolving paradigms of manufacturing: from mass production to mass customization and personalization. *Procedia CIRP* 2013;7:3–8.
23. Balakrishnan A, Kumara SR, Sundaresan S. Manufacturing in the digital age: exploiting information technologies for product realization. *Inf Syst Front* 1999;1(1):25–50.
24. Bughin J, Chui M, Manyika J. Clouds, big data, and smart assets: ten tech-enabled business trends to watch. *McKinsey Q* 2010;56(1):75–86.
25. Li J, Tao F, Cheng Y, Zhao L. Big data in product lifecycle management. *Int J Adv Manuf Technol* 2015;81(1–4):667–84.
26. Karthigadevi K, Srinivasagan G. Random Forest classification algorithm for agricultural data analysis in Tirunelveli district. *J Xi'an Univ Archit Technol* 2020;12:418–32.
27. Lee J, Lapira E, Bagheri B, Kao HA. Recent advances and trends in predictive manufacturing systems in big data environment. *Manuf Lett* 2013;1(1):38–41.
28. Siddiqa A, Hashem IAT, Yaqoob I, Marjani M, Shamshirband S, Gani A, et al. A survey of big data management: taxonomy and state-of-the-art. *J Network Comput Appl* 2016;71:151–66.
29. Zhang Y, Zhang G, Wang J, Sun S, Si S, Yang T. Real-time information capturing and integration framework of the internet of manufacturing things. *Int J Comput Integr Manuf* 2015;28(8):811–22.
30. Tao F, Cheng J, Qi Q. IIHub: an industrial Internet-of-Things hub towards smart manufacturing based on cyber–physical system. *IEEE Trans Ind Inf* 2017, http://dx.doi.org/10.1109/TII.2017.2759178
31. Guerriero A, Ragni F, Martines C. A dynamic URL assignment method for parallel web crawler. In: *Proceedings of 2010 IEEE International Conference on Computational Intelligence for Measurement Systems and Applications* (pp. 119–23). 2010. IEEE.

32. Gandomi A, Haider M. Beyond the hype: big data concepts, methods, and analytics. *Int J Inf Manage* 2015;35(2):137–44.
33. Nicolae B, Antoniu G, Bougé L, Moise D, Carpen-Amarie A. BlobSeer: next-generation data management for large scale infrastructures. *J Parallel Distrib Comput* 2011;71(2):169–84.
34. Agrawal D, El Abbadi A, Antony S, Das S. Data management challenges in cloud computing infrastructures. In S. Kikuchi, S. Sachdeva, S. Bhalla, editors. *Databases in Networked Information Systems: 6th International Workshop, DNIS 2010, Aizu-Wakamatsu, Japan, March 29–31, 2010* (Lecture Notes in Computer Science, vol. 5999, pp. 1–10). 2010. Springer.
35. Huang J, Li YF, Xie M. An empirical analysis of data preprocessing for machine learning-based software cost estimation. *Inf Software Technol* 2015;67:108–27.
36. Begoli E, Horey J. Design principles for effective knowledge discovery from big data. In: *Proceedings of the Joint 10th Working IEEE/IFIP Conference on Software Architecture & 6th European Conference on Software Architecture* (pp. 215–18). 2012. IEEE.
37. Mittal S, Khan MA, Romero D, Wuest T. Smart manufacturing: characteristics, technologies and enabling factors. Proceedings of the Institution of Mechanical Engineers. *Part B: J Eng Manuf* 2017, http://dx.doi.org/10.1177/0954405417736547
38. Tao F, Cheng J, Qi Q, Zhang M, Zhang H, Sui F. Digital twin-driven product design, manufacturing and service with big data. *Int J Adv Manuf Technol* 2017;94(9–12):3563–76. https://doi.org/10.1007/s00170-017-0233-1
39. Liu Y, Jin J, Ji P, Harding A, Fung RY. Identifying helpful online reviews: a product designer's perspective. *Comput Aided Des* 2013;45(2):180–94.
40. Bennane A, Yacout S. LAD-CBM; new data processing tool for diagnosis and prognosis in condition-based maintenance. *J Intell Manuf* 2012;23(2):265–75.
41. Kusiak A, Verma A. Analyzing bearing faults in wind turbines: a data-mining approach. *Renew Energ* 2012;48:110–6.
42. Karthigadevi K, Sakthimohan M, Tamilselvan S, Thiyagarajan B, Elizabeth Rani G, Sanjeevi Kumar V. Enhanced 16 × 16 Dadda multiplier using Kung–Brent Adder for superior parallel processing performance. In: *Proceedings of the 8th International Conference on Electronics, Communication and Aerospace Technology (ICECA), Coimbatore, India* (pp. 80–85). 2024,
43. Qi J, Zhang Z, Jeon S, Zhou Y. Mining customer requirements from online reviews: a product improvement perspective. *Inf Manage* 2016;53(8):951–63.
44. Karuthapandi S, Murugan R. An experimental investigation of flat wire electrodes and their weld bead quality in the FCAW process. *High Temp Mater Process* 2017;21(1):65–79.
45. Innovation Kusiak A. A data-driven approach. *Int J Prod Econ* 2009;122(1):440–8.
46. Da Cunha C, Agard B, Kusiak A. Selection of modules for mass customisation. *Int J Prod Res* 2010;48(5):1439–54.
47. Wu D, Liu X, Hebert S, Gentzsch W, Terpenny J. Democratizing digital design and manufacturing using high performance cloud computing: performance evaluation and benchmarking. *J Manuf Syst* 2017;43:316–26.

48. Ji W, Wang L. Big data analytics based fault prediction for shop floor scheduling. *J Manuf Syst* 2017;43:187–94.
49. Cheng Y, Tao F, Zhang L, Zhao D. Dynamic supply-demand matching for manufacturing resource services in service-oriented manufacturing systems: a hypernetwork-based solution framework. In: *Proceedings of ASME International Manufacturing Science and Engineering Conference (MSEC2015), Charlotte, NC, USA, June 8–12*, 2015.
50. Song Z, Kusiak A. Multiobjective optimization of temporal processes. *IEEE Trans Syst Man Cybern Part B (Cybern)* 2010;40(3):845–56.
51. Tao F, Zhao D, Hu Y, Zhou Z. Resource service composition and its optimal-selection based on particle swarm optimization in manufacturing grid system. *IEEE Trans Ind Inf* 2008;4(4):315–27.
52. Anandavalli R, Karthigadevi K, Elizabeth Rani G, Sakthimohan M, Srinivasan G. A flutter app for AI-based plant disease detection. In: *Proceedings of the 8th International Conference on Electronics, Communication and Aerospace Technology (ICECA)*, Coimbatore, India, 2024, pp. 661–666. doi: 10.1109/ICECA63461.2024.10801001
53. Zhong RY, Huang GQ, Lan S, Dai QY, Chen X, Zhang T. A big data approach for logistics trajectory discovery from RFID-enabled production data. *Int J Prod Econ* 2015;165:260–72.
54. Lu S, Xu C, Zhong RY, Wang L. A RFID-enabled positioning system in automated guided vehicle for smart factories. *J Manuf Syst* 2017;44:179–90.
55. Qin SJ. Survey on data-driven industrial process monitoring and diagnosis. *Annu Rev Control* 2012;36(2):220–34.
56. Qin SJ. Process data analytics in the era of big data. *AIChE J* 2014;60(9):3092–100.
57. Abonyi J, Roubos JA, Szeifert F. Data-driven generation of compact: accurate, and linguistically sound fuzzy classifiers based on a decision-tree initialization. *Int J Approximate Reasoning* 2003;32(1):1–21.
58. Karthigadevi K, Balamurali S, Venkatesulu M. Improving quality of service in wireless sensor networks using neighbor constraint transmission centric distributed sink hole detection and network simulator 2. *ARPN J Eng Appl Sci* 2017;12(4):1197–201.
59. Kim A, Oh K, Jung JY, Kim B. Imbalanced classification of manufacturing quality conditions using cost-sensitive decision tree ensembles. *Int J Comput Integr Manuf* 2017;31(8):701–17, http://dx.doi.org/10.1080/0951192X.2017.1407447
60. Köksal G, Batmaz I, Testik MC. A review of data mining applications for quality improvement in manufacturing industry. *Expert Syst Appl* 2011;38(10):13448–67.
61. Sata A, Ravi B. Bayesian inference-based investment-casting defect analysis system for industrial application. *Int J Adv Manuf Technol* 2017;90(9–12):3301–15.

Chapter 8

Additive manufacturing applications in automotive and aviation industries

K. Vignesh

8.1 INTRODUCTION

Additive manufacturing (AM), also known as three-dimensional (3D) printing, is an emerging technology attracted by researchers and modern industries because it can reduce the number of processes to fabricate lightweight products without compromising the required performance and structural integrity. The lightweight component is essential for various industries but is not limited to aerospace, automotive, biomedical, and other sectors. One of the main advantages of this technology is that intricate shapes can be easily manufactured without any special tools. Conventional manufacturing is used to subtract unnecessary materials from the raw materials which can increase the waste, whereas AM is used to fabricate the part by adding one layer over another according to the details provided in the computer-aided design (CAD) design and can reduce the waste. According to the American Society for Testing and Materials (ASTM-F42), the various technologies of AM are classified into seven major groups as shown in Figure 8.1 and the raw materials used for this process may be in any form, such as solid (wire or filament), powder (metal or polymer fine granules), and liquid (liquid polymers, resins, and photopolymers). The AM can produce components using a variety of technical materials, including ceramics, polymers, composites, and metals for complex designs and intricate shapes [1].

Among the various AM technologies, as shown in Figure 8.1 , the preferable AM technology for the aviation sector is electron beam melting (EBM), direct metal laser sintering (DMLS), selective laser melting (SLM), and wire arc additive manufacturing (WAAM) [2]. Table 8.1 presents the benefits and limitations of the AM processes. The aerospace industry held 19% of the AM market in 2017 which rose to 23% in 2021. It is predicted that, this industry will hold 25% of the total market at the end of 2025 (Figure 8.2) [3]. The AM product for the aerospace sector is not only preferred for lightweight components due to its sustainability in high-temperature environments. For high-temperature aviation applications, materials like nickel alloys, titanium alloys, etc. can be used to produce intricate shapes using the

DOI: 10.1201/9781003569633-8

ASTM CLASSIFICATION	RAW MATERIAL	AM PROCESS
(a) Vat Photopolymerization	(a) Liquid	(a) SLA, DLP, LCD
(b) Material Jetting	(b) Liquid	(b) Material Jetting
(c) Sheet Lamination	(c) Solid	(c) Sheet Lamination
(d) Material Extrusion	(d) Solid	(d) FDM or FFF
(e) Directed Energy Deposition	(e) Solid	(e) LMD. WAAM, L-DED
(f) Binder Jetting	(f) Powder	(f) Binder Jetting
(g) Powder Bed Fusion	(g) Powder	(g) SLS, EBM, DMLS or SLM

Figure 8.1 Seven groups.

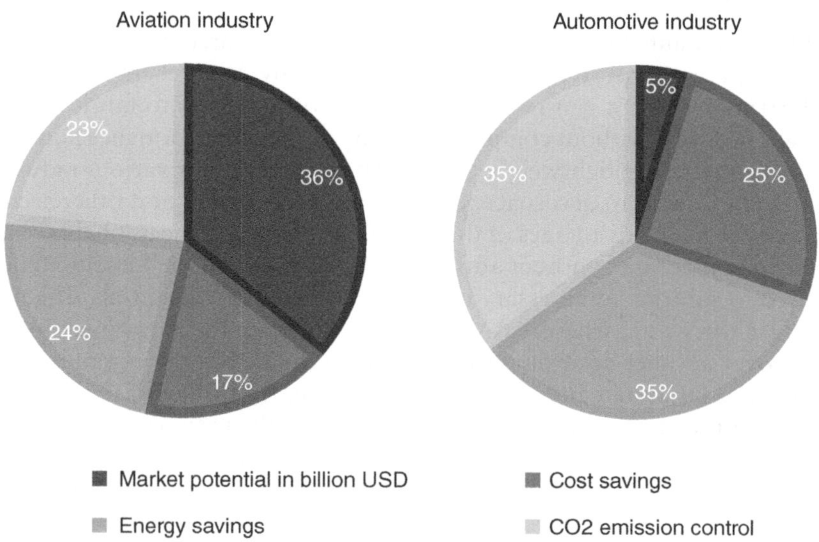

Figure 8.2 Total market.

Table 8.1 Benefits and limitations of the AM process

Benefits	*Limitations*
Final parts with variable infill density	Post processing is required
Buy-to-fly ratio 1:1	Not suitable for mass production
Can alternate the design for every component	Not suitable for large size of the component
Lightweight components can be produced.	High initial investment
Suitable for batch production	Slow fabrication process
Complex and intricate shapes can be made easily	Limited to few materials
No complex tooling required	Not globally standardized
Minimum material waste	Defects like hot cracking and porosity.

AM process, which is very difficult to produce with conventional methods like casting and machining.

8.2 STEPS INVOLVED IN AM

AM is the process of producing a real-time component from CAD data. Several steps are involved to get the final product.

8.2.1 CAD design and exporting

The initial step in the AM process is to design the model using CAD software. Some of the CAD software, such as AutoCAD, CATIA, SolidWorks, Fusion360, and ANSYS, are used for designing the model for AM [4–8]. After their design, it is crucial to export the models to the appropriate format. The exporting file formats are STL, STEP, IGES, AMF, 3MF, PLY, FBX, Collada, X3D, and SVG. Of these, some of the formats like SVG can be used for laser cutting [9].

8.2.2 Slicing

The slicing operation is the next step followed by the design and modeling. In this step, the exported file is given as the input to the slicing software. Several commercial and open access software are available in the market. The slicing operation will decide the layer thickness, printing speed, bed temperature, and nozzle size. This step also decides the support structure requirement and path movement of the nozzle.

8.2.3 Printing and post processing

The setting of the printer, like during leveling the table, feeding the correct material as given during slicing, and cleaning the printing bed, is essential before printing. During printing, it is required to ensure that the feedstock is continuously moving without any disturbances, because the winding of the feedstock crosses over the other. After printing post processing is required. For instance, the WAAM-produced components need machining after the printing.

8.3 AM TECHNOLOGIES IN AVIATION

As discussed earlier, all AM processes are categorized into seven major categories.

The selection of processes is based on the materials used, the application, and environmental aspects. All the AM processes can produce a lightweight model as compared to conventional manufacturing. Therefore, the AM process in aviation can save fuel, reduce emissions, increase the payload, and improve the performance of the aircraft.

8.3.1 Fused filament fabrication (FFF)

In FFF the filament is fused to fabricate the 3D component [10]. This process is also known as fused deposition modeling (FDM) [11]. The FFF machine is the lowest price printer in the AM market. The thermoplastic polymers as solid filament is used as the raw material in this process. Generally, the filament size is 1.75 mm, and the extruder temperature is a maximum of around 400°C. Complex geometries with good surface finish can be attained using this process which is preferred in aircraft industries [12, 13]. In total, the FFF is transforming the aviation industry with better precision and reliability. The FFF technology has one of the major limitations, which is the size of the product. The component size in FFF is dependent on the bed size of the printer. For large-size parts, a larger printer bed is required. However, the FFF is becoming a reliable source of technology in aviation sectors including the models of prototypes before the final fabrication model which helps understand the behavior of the model, the assembly and fabrication tools like jigs and fixtures, the air conditioning system in aircraft uses duct with complex profiles, external parts of an engine like housings and brackets, the components such as tray table, overhead bins and armrests used for interior design, landing gear, frame, and wings for the components of unmanned aerial vehicle (UAV) can be manufactured using the FFF technology.

8.3.2 Selective laser melting/sintering

In SLM or selective laser sintering (SLS) technology, laser is used as the power source and the raw material is in powder form. In general, SLM is used for metals and SLS is used for polymers. Using SLM/SLS technology, producing the components is popular in aviation industry because it can fabricate intricate and complex geometries with good accuracy and surface finish. The cost of the machine, processing cost, and price of the powders are high. However, this technology is a highly preferable process in aviation industries because it can fabricate the product with high melting point powders with better dimensional accuracy and lightweight components with high strength. The SLS/SLM components like engine parts, ducts, turbine blades, and brackets are used in aircraft and satellites [14, 15].

8.3.3 Electron beam melting

In EBM, the powder is used as a raw material similar to SLM/SLS, but the EBM is done under a vacuum only. This process requires high cost and produces a good surface finish, better dimensional accuracy. The size of the component is limited due to the vacuum area. The EBM components are highly used in satellites and aircraft.

8.3.4 Wire arc additive manufacturing

WAAM is the arc welding technology used to fabricate 3D components. In the metal AM techniques, the WAAM is the only process that has the highest deposition rate of 1–10 kg per hour [16] and also used for repairing of components [17]. The advantages of WAAM are large-sized components that can be fabricated, the fastest building rate, and applicability to all materials like stainless steel, super alloys, copper alloys, brass alloys, and aluminum alloys. Solid filament wire is used as the feedstock for WAAM. The cost of the feedstock is very low as compared to the powder of the same material. The cost of the wire is nearly 10% of the cost of the powder for the same material and the same weight [18]. These are some of the types of WAAM: (a) gas metal arc welding (GMAW), (b) gas tungsten arc welding (GTAW), and (c) plasma arc welding (PAW).

8.3.5 Post processing

AM manufacturing is a developing and reliable technology in aviation sector. After fabricating the final product, some components required post processing according to the technology adopted. The components fabricated by similar or dissimilar materials may complicate the post-processing procedures. Apart from this, post processing may affect the surface finish, dimensional tolerance, removal of support structure, and structural integrity. The post-processing techniques like chemical treatments, polishing, and sanding may reduce the adverse effects. Due to the size limitation, large parts require bonding and the post processing is essential in the bonding areas to ensure good bonding strength. The component to be used in aircraft must pass stringent aircraft standards which ensure the AM part quality in terms of strength-to-weight ratio, dimensional accuracy, fatigue strength, and structural integrity. Therefore, the effective post-processing techniques can help the AM-fabricated parts to be used in aircraft and UAVs [15, 19, 20].

8.4 MATERIALS USED IN AM PROCESSES

In any process, the selection of material is an important task. Based on the application the materials will be selected. A variety of engineering materials are available; they are aluminum and their alloys, copper and its alloys, stainless steel and alloys, magnesium and alloys, alloys of copper and bronze, nickel based super alloys, iron-based superalloys, polymers, ceramics, and other materials. The raw materials may be in any form according to the selection of the AM process. In aerospace for a long time, aluminum and its alloys have been widely used because this material has high strength at low weight, low cost, and is easy to manufacture. However, due to the advancement in material research, several composites and superalloys have come into picture [21, 22]. However, the alloys of aluminum are slightly reduced in the aircraft because they did not perform well at higher temperatures and had poor resistance to corrosion [23]. Stainless steel is one of the alternatives to resolve this problem because, at elevated temperatures, it will perform well with better mechanical characteristics, good corrosion resistance, good durability, and resistance to oxidation [24]. The stainless-steel classifications like precipitation hardened steel, maraging steel, and austenitic stainless steel are used as aircraft components such as heat exchangers, exhaust components, fasteners for joints, parts of an engine, and landing gears.

Another alternative to aluminum and steel is Ti alloys, which possess a high strength-to-weight ratio, excellent temperature resistance at elevated temperatures, and better corrosion resistance [25]. Due to the admirable properties of titanium alloys, it is widely used in aircraft and satellites [26]. Recently, carbon-fiber-reinforced polymer (CFRP) has been combined with Ti alloys and is used in new model aircraft and the cryogenic propellant tank carrying cryogenic fuels [27, 28]. Therefore, Ti alloys are useful for high-temperature and low-temperature applications. Superalloys are becoming popular for high-temperature and high-pressure applications [29]. The superalloys can be classified based on iron or nickel. These superalloys can be found in several cryogenic and high-temperature applications like components of high-pressure turbine engines [30, 31] and injectors, manifolds, valves, and igniters [32] because the properties of these alloys do not degrade at high pressure, temperature, and corrosive environments.

8.5 APPLICATIONS OF AM IN AVIATION INDUSTRIES

In the aviation industry, AM plays a crucial role in fuel efficiency by producing a lightweight component with complex shapes such as fuel nozzles, turbine blades, and compressors [33, 34]. The lightweight is a true advantage for AM technology, but this can be achieved only by optimal design consideration. The aviation performance and running cost can be improved by optimizing the structures like lattice, cellular, and honeycomb [35, 36]. Some of the components in aircraft are tough to fabricate as single components with multiple functions because it raises the functional complexity. For instance, one of the crucial components in the aircraft engine is a swirler which is used to generate a high turbulent flow by forcedly recirculating the combustion fuel inside the chamber. The chamber pressure will be reduced when the air turbulence increases. Therefore, the design of swirler geometry must be carefully done to allow the anticipated turbulence to mix the pressurized fuel inside the chamber and airflow. The swirler manufactured using the AM technique reduces the functional complexity and increases the flight performance [37].

8.5.1 Aircraft applications

The SLM process is also known as direct metal laser sintering. The A380 Airbus was the first flight that used additively manufactured hydraulic components and flew in the air [38]. The aircraft model A320 is owned by an Airbus innovation group, and they achieved a 64% weight reduction in the nacelle hinge bracket in the DMLS AM process after the optimal design implementation [39]. This reduced weight not only increased fuel efficiency but also reduced carbon emission by 40% [40]. The original part was manufactured by cast steel in conventional manufacturing and the optimized part was manufactured with the material titanium [41]. The Airbus model A320neo MTU aero engines utilized the DMLS AM process to fabricate the optimized nickel alloy borescope bosses and reduced the consumption of fuel by 15% [42]. The borescope is the component mounted on the turbine to check the turbine blade condition inside the engine. The rocket injector of the Ariane 6 rocket had the assembly of 248 single components. The AM process fabricated the same injector as the single component and reduced the weight [43]. SpaceX, a private sector player in aerospace, utilized the SLM AM process and produced the Inconel-based superDraco engines for the capsule of Dragon V2 [44]. The Airbus of A350 XWB used the SLM process to fabricate the first porous titanium bracket connector (Figure 8.3) in the market which reduced the weight by 30% and process time by 75%. With the help of this technology, Airbus introduced the pilotless aircraft or UAVs known as THOR in the aviation market [45]. During the bracket design and construction for aircraft and defense sector applications, the engineers of Airbus encountered some problems and the two major problems were ensuring robust connection and enduring at high-temperature environments during flying. Therefore, the engineers chose 3D printing or AM technology to print the bracket and the problems were solved [46].

The SLS/DMLS AM process has a good surface finish, but it is limited by size and has high equipment costs. Another AM process called WAAM can overcome these limitations. Apart from the above limitations, the porosity and wing spar are the aircraft wing's skeleton, also known as the backbone of the wing. Cranfield University and BAE Systems used the WAAM process to manufacture a titanium wing spar component with a large length of 1.2 m in the open environment [47, 48].

8.5.2 AM standards for aviation

Despite several advantages of AM, only a few AM-fabricated components are allowed to be used in the aviation industry due to the strict certification

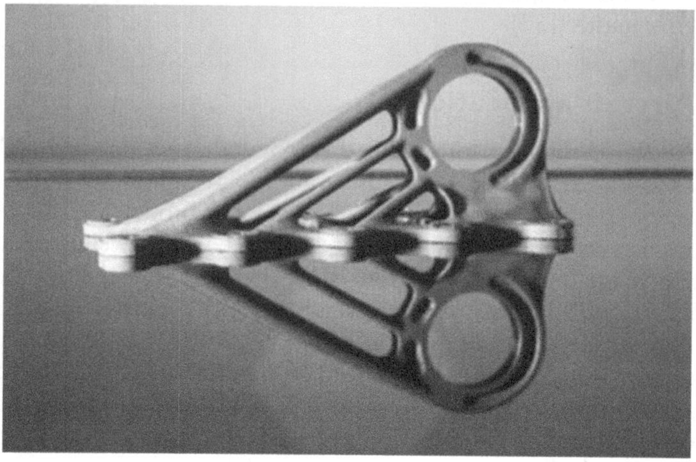

Figure 8.3 Airbus.

Table 8.2 Standards currently in use for the AM process in aviation industries

Standard	*Purpose/description*
ISO/ASTM52900:21	General principle of AM, fundamentals and vocabulary
ASTM F3049-14 (2021)	Metal powder characterization used in the AM process
MSFC-STD-3716 (NASA)	LBPF AM parts for flight applications demanding high reliability.
AS9100D	Space, defense, and aviation industries require Quality Management Systems (QMS).
ISO 9001:2015	Globally recognized QMS.

rules and lack of established standards [49]. Regulatory bodies of aviation and aviation manufacturers are working toward establishing the standards to resolve the contradiction between industrial sectors and academic institutions. Globally, organizations like ISO (International Organization for Standardization), ASTM (American Society for Testing and Materials), MSFC (Marshall Space Flight Center) established by NASA (National Aeronautics and Space Administration), and AS (Aerospace Standard) released by IAQG (International Aerospace Quality Group) are working toward the AM standards. Some of the available standards for metal AM parts are presented in Table 8.2.

8.6 APPLICATIONS OF AM IN AUTOMOTIVE INDUSTRIES

The automobile industry influences the global economy. In the year 2030, it is expected that the automotive market globally will reach around

3,80,0000 million US dollars and it is estimated that the economy is growing at a compound annual growth rate (CAGR) of 3%. There are several areas the automobile industries need to pay attention to improve the global economy. The areas are environmental impact solutions like eco-friendliness, research on fully autonomous and electric vehicles (EVs), solutions to reach transport facilities to the interior of the villages, and customized design vehicles. Globally, the automobile sector is more innovative and producing products apart from conventional internal combustion engines and compression ignition engines such as EVs, hybrid EVs, hydrogen-based engines on research, and Internet of Things (IoT)-based automobiles.

Apart from the various innovations and directions, as mentioned above, manufacturing advancement is also an important criterion for reducing fuel consumption and carbon emissions. Advanced manufacturing techniques like AM are helpful in producing lightweight components with increased strength [50]. Moreover, still, the industries are still reliable in casting, forming, and machining processes for mass production. However, adopting the AM process in the manufacturing unit can increase the buy-to-fly ratio as compared to the subtractive manufacturing [51]. The AM components in the automobile market will hit 2.5–7.9 billion US dollars from the year 2022 to 2027 at a CAGR of around 25% as reported.

8.6.1 Applications of AM in automotive industries

According to the trend, the automotive industries are developing new models with the evolution of technology in all aspects like safety, aesthetics, lightweight, and fuel efficiency aerodynamics [52]. The AM process can be used to address the above challenges in the automotive industry. The automobile industry uses various AM processes like binder jetting, sheet lamination, powder bed fusion, material extrusion, and direct energy deposition (DED) [53]. Plastics, metals, and their alloys and ceramics are the materials used for fabricating the product using the AM process in automobile industries [54]. AM is preferred in automobile industries because it can fabricate lightweight components without affecting their strength and functionality [55].

The crankshaft in the engine plays an important role in converting rotary to reciprocating motion. The designers in the Honda industries and Autodesk designed and developed the crankshaft CAD model using Fusion 360 and Netfabb [56]. The fabricated crankshaft used AM process and reduced 50% of the component weight [57]. General Motors replaced the component assembly of eight pieces into AM-fabricated single component [58]. Around 1986, Ford invested in AM technology and the first-ever SLA technology was utilized by them [59]. Ford manufactures various parts of an automobile continuously by using various AM technologies like DED, SLA, DMLS, etc. in the advanced manufacturing at Detroit, USA. Volkswagen, one of the popular automobile companies, has been using AM technology for the past 25 years

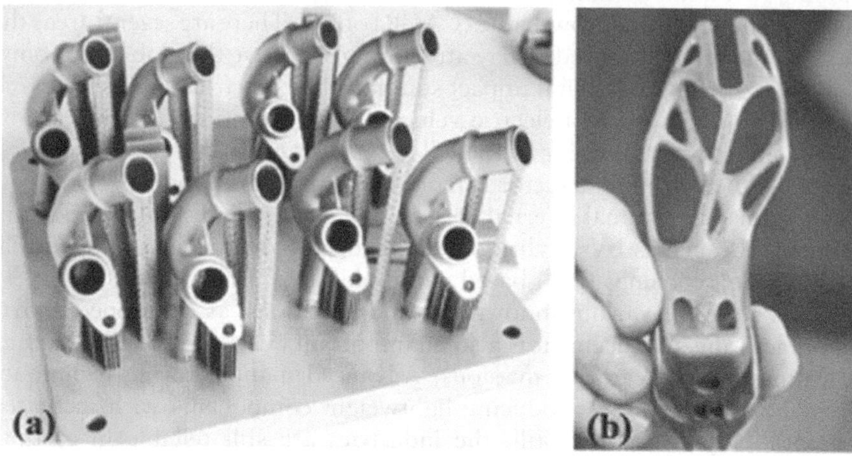

Figure 8.4 W12 and i8.

[60]. In 1991, BMW first printed the components using the SLA process, and now an average of 1 lakh AM components are fabricated every year. The BMW celebrated its 25th anniversary of 3D printing in the year 2015 [61]. BMW has invested in AM technology of approximately 15 million Euros to set up the AM Campus in Munich, Germany in the year 2020. Several parts like license plates, bumpers, and a few prototypes of BMW cars like the BTM race car and BMW i8 roadster are manufactured using AM technology [62, 63]. This indicates that AM is not new for the automotive industry. A German company Electro Optical System (EOS) GmbH tested a new material Al2139 aluminum alloy in their vehicle due to its high temperature resistance, tensile strength of 500 MPa, and good corrosion resistance. They chose the AM technique to produce the component with a time saving of 88% compared to conventional manufacturing [53]. Figure 8.4 shows the topology optimized water connector for Audi engine model W12 and the fixture used for the i8 Roadster soft-top [64]. The largest DMLD part printed with the help of AM technology was used in the engine manifold of the 1977 Hoonitruk [65, 66]. The engine manifold was made up of aluminum and the engine cylinders received an air-fuel mixture through this manifold. Approximately 5 days were taken to fabricate the largest manifold.

8.7 CONCLUSION

The AM technology is used to reduce the weight of the product but not sacrificing its strength. Topology optimization helps achieve a strength-to-weight ratio. Lightweight materials with high strength have higher costs than steel like titanium alloys. Titanium alloys cost approximately five times more than aluminum alloys. The AM process can reduce the usage of material and achieve the properties of the original product as manufactured by using conventional methods. However, in-depth research is required to pass the AM standards and certifications of aerospace. Also, there is a need to develop more standards and certifications for aerospace according to technology development.

DMLS technology is popularly accepted in aviation sectors and DED is very few. The advantage of DED is that it can be used for repairing and can produce large parts with a high deposition rate. DMLS can apply only for small to medium-sized components. Even with a lot of merits in AM technology, the components used in aircraft are limited because of distinct requirements for quality control, low rate of production, challenges in post processing, and high initial investment.

The automobile industry has been using the AM process since the early 1990s. All automobile industries start their AM journey with the SLA process and invest a lot of money in constructing AM labs and future research. Materials like plastics, steel, aluminum, titanium, and magnesium are used in the AM process. The automotive industry is using several AM techniques like FDM, WAAM, SLS, and binder jetting. At present, automobiles are being upgraded to EVs and several AM parts are used in EVs to reduce the weight of the vehicle.

The AM process needs to be studied in detail for further implementation of AM process in aviation and automotive sectors. The future research's important topics are listed below.

- Implementation of AI models like a digital twin for online monitoring quality control and identifying the flaws during the process itself.
- Attention to lattice structures.
- Effective use of optimization models like topology optimization.
- Research on the development of new alloys and composites with lightweight, high strength, and moderate cost.

REFERENCES

1. Vaezi M, Chianrabutra S, Mellor B, Yang S. Multiple material additive manufacturing—part 1: A review. *Virtual and Physical Prototyping* 2013;8:19–50. https://doi.org/10.1080/17452759.2013.778175.
2. Radhika C, Shanmugam R, Ramoni M, Gnanavel BK. A review on additive manufacturing for aerospace application. *Materials Research Express* 2024;11:022001. https://doi.org/10.1088/2053-1591/ad21ad.
3. Gebler M, Schoot Uiterkamp AJM, Visser C. A global sustainability perspective on 3D printing technologies. *Energy Policy* 2014;74:158–67. https://doi.org/10.1016/j.enpol.2014.08.033.

4. Albu SC. Development on the production of a new type of extruder used in additive manufacturing, FDM technology. *Procedia Manufacturing* 2018;22:141–6. https://doi.org/10.1016/j.promfg.2018.03.022.
5. Krishnaraj J, Nidumolu ECP. Modeling and additive manufacturing of a scaled four cylinder petrol engine. *AIP Conference Proceedings* 2023;2492:040056. https://doi.org/10.1063/5.0113221.
6. Rosso S, Savio G, Uriati F, Meneghello R, Concheri G. Optimization approaches in design for additive manufacturing. *Proceedings of the Design Society: International Conference on Engineering Design* 2019;1:809–18. https://doi.org/10.1017/dsi.2019.85.
7. Patil SD, Desai AA, Hatte PR. Usage of autodesk fusion 360 software for 3D reconstruction of anatomical structure from the 2D X-ray image. *Medical Additive Manufacturing*, Elsevier; 2024, pp. 419–41. https://doi.org/10.1016/B978-0-323-95383-2.00020-2.
8. Jankovics D, Gohari H, Tayefeh M, Barari A. Developing topology optimization with additive manufacturing constraints in ANSYS®. *IFAC-PapersOnLine* 2018;51:1359–64. https://doi.org/10.1016/j.ifacol.2018.08.340.
9. Toxirjonovich ON, o'g'li DMK. 3D modeling and computer graphics in virtual reality. *International Journal on Orange Technologies* 2021;3:61–6.
10. Krishnanand, Soni S, Taufik M. Design and assembly of fused filament fabrication (FFF) 3D printers. *Materials Today: Proceedings* 2021;46:5233–41. https://doi.org/10.1016/j.matpr.2020.08.627.
11. Krishnanand, Taufik M. Fused filament fabrication (FFF) based 3D printer and its design: A review. Advanced Manufacturing Systems and Innovative Product Design. Lecture Notes in Mechanical Engineering, Springer; 2021, pp. 497–505. https://doi.org/10.1007/978-981-15-9853-1_41.
12. Skawiński I, Goetzendorf-Grabowski T. FDM 3D printing method utility assessment in small RC aircraft design. *Aircraft Engineering and Aerospace Technology* 2019;91:865–72. https://doi.org/10.1108/AEAT-07-2018-0189.
13. Taylor RM, Niakin B, Lira N, Sabine G, Lee J, Conklin C, et al. Design optimization, fabrication, and testing of a 3D printed aircraft structure using fused deposition modeling. AIAA Scitech 2020 Forum, American Institute of Aeronautics and Astronautics; 2020. https://doi.org/10.2514/6.2020-1924.
14. Samal SK, Vishwanatha HM, Saxena KK, Behera A, Nguyen TA, Behera A, et al. 3D-printed satellite brackets: Materials, manufacturing and applications. *Crystals (Basel)* 2022;12:1148. https://doi.org/10.3390/cryst12081148.
15. Subeshan B, Ali Z, Asmatulu E. Metal additive manufacturing in space and aerospace exploration: Current processes, material selection and challenges. *Journal of Engineering and Applied Sciences* 2024;11:35. https://doi.org/10.5455/jeas.2024021104.
16. Alagha AN, Hussain S, Zaki W. Additive manufacturing of shape memory alloys: A review with emphasis on powder bed systems. *Materials and Design* 2021;204:109654. https://doi.org/10.1016/j.matdes.2021.109654.
17. Lee JH, Lee CM, Kim DH. Repair of damaged parts using wire arc additive manufacturing in machine tools. *Journal of Materials Research and Technology* 2022;16:13–24. https://doi.org/10.1016/j.jmrt.2021.11.156.

18. Shah A, Aliyev R, Zeidler H, Krinke S. A review of the recent developments and challenges in wire arc additive manufacturing (WAAM) process. *Journal of Manufacturing and Materials Processing* 2023;7:97. https://doi.org/10.3390/jmmp7030097.
19. Hao P, Li Y, Huan D, Wang Z, Xiao J. Exploration of process parameters and post-processing of 3D printed SCF/PPS for low-altitude aircraft fabrication. *Aerospace Traffic and Safety* 2025;2:10–19. https://doi.org/10.1016/j.aets.2024.12.008.
20. Patel KN, Manoj A, Sayed MSH, Shah Kaushal K, Shah SJ, Dave HK. Post processing 3D printed UAV wing enabling trailing edge morphing technology. Recent Advances in Manufacturing Processes and Systems. Lecture Notes in Mechanical Engineering, Springer; 2022, pp. 791–802. https://doi.org/10.1007/978-981-16-7787-8_63.
21. Parveez B, Kittur MI, Badruddin IA, Kamangar S, Hussien M, Umarfarooq MA. Scientific advancements in composite materials for aircraft applications: A review. *Polymers (Basel)* 2022;14:5007. https://doi.org/10.3390/polym14225007.
22. Zhang X, Chen Y, Hu J. Recent advances in the development of aerospace materials. *Progress in Aerospace Sciences* 2018;97:22–34. https://doi.org/10.1016/j.paerosci.2018.01.001.
23. Dwivedi P, Siddiquee AN, Maheshwari S. Issues and requirements for aluminum alloys used in aircraft components: State of the Art. *Russian Journal of Non-Ferrous Metals* 2021;62:212–25. https://doi.org/10.3103/S1067821221020048.
24. Morris Jr JW. Making steel strong and cheap. *Nature Materials* 2017;16:787–9. https://doi.org/10.1038/nmat4949.
25. Kishawy HA, Hosseini A. Titanium and titanium alloys. Machining Difficult-to-Cut Materials. Materials Forming, Machining and Tribology, Springer; 2019, pp. 55–96. https://doi.org/10.1007/978-3-319-95966-5_3.
26. Anil Kumar V, Gupta RK, Prasad MJNV, Narayana Murty SVS. Recent advances in processing of titanium alloys and titanium aluminides for space applications: A review. *Journal of Materials Research* 2021;36:689–716. https://doi.org/10.1557/s43578-021-00104-w.
27. Kayihan M, Karaguzel U, Bakkal M. Process design and experimental study on drilling operations of a hybrid aluminum/carbon fiber reinforced polymer/titanium composite. *Materials and Manufacturing Processes* 2024;39:1630–7. https://doi.org/10.1080/10426914.2024.2368547.
28. Ogasawara T, Arai N, Fukumoto R, Ogawa T, Yokozeki T, Yoshimura A. Titanium alloy foil-inserted carbon fiber/epoxy composites for cryogenic propellant tank application. *Advanced Composite Materials* 2014;23:129–49. https://doi.org/10.1080/09243046.2013.844756.
29. Prashar G, Thakur K, Singh S, Singh P, Srivastava VK. Superalloys for high-temperature applications: An overview. AIP Conference Proceedings; 2024, p. 020022. https://doi.org/10.1063/5.0192482.
30. Biswas S, Ramachandra S, Hans P, Kumar SPS. Materials for gas turbine engines: Present status, future trends and indigenous efforts. *Journal of Indian Institute of Science* 2022;102:297–309. https://doi.org/10.1007/s41745-022-00295-z.

31. Gudivada G, Pandey AK. Recent developments in nickel-based superalloys for gas turbine applications: Review. *Journal of Alloys and Compounds* 2023;963:171128. https://doi.org/10.1016/j.jallcom.2023.171128.
32. Salunkhe S, Rajamani D. Current trends of metal additive manufacturing in the defense, automobile, and aerospace industries. Advances in Metal Additive Manufacturing, Elsevier; 2023, pp. 147–60. https://doi.org/10.1016/B978-0-323-91230-3.00004-4.
33. Madhavadas V, Srivastava D, Chadha U, Aravind Raj S, Sultan MTH, Shahar FS, et al. A review on metal additive manufacturing for intricately shaped aerospace components. *CIRP Journal of Manufacturing Science and Technology* 2022;39:18–36. https://doi.org/10.1016/j.cirpj.2022.07.005.
34. Liu R, Wang Z, Sparks T, Liou F, Newkirk J. Aerospace applications of laser additive manufacturing. Laser Additive Manufacturing, Elsevier; 2017, pp. 351–71. https://doi.org/10.1016/B978-0-08-100433-3.00013-0.
35. Pan C, Han Y, Lu J. Design and optimization of lattice structures: A review. *Applied Sciences* 2020;10:6374. https://doi.org/10.3390/app10186374.
36. Karuthapandi S, Murugan R, Palani PK. Study and analysis of the macrostructure characteristics in FCAW with the use of a flat wire electrode and by optimizing the process parameter using the Taguchi method and regression analysis. *High Temperature Material Processes* 2016;20(3):197–224.
37. Karuthapandi S, Ramu M, Thyla PR. Effects of the use of a flat wire electrode in gas metal arc welding and fuzzy logic model for the prediction of weldment shape profile. *Journal of Mechanical Science and Technology* 2017;31(5):2477–86.
38. Wu W, Xia R, Qian G, Liu Z, Razavi N, Berto F, et al. Mechanostructures: Rational mechanical design, fabrication, performance evaluation, and industrial application of advanced structures. *Progress in Materials Science* 2023;131:101021. https://doi.org/10.1016/j.pmatsci.2022.101021.
39. Giuliani F, Paulitsch N, Cozzi D, Görtler M, Andracher L. An assessment on the benefits of additive manufacturing regarding new swirler geometries for gas turbine burners. Volume 4A: Combustion, Fuels, and Emissions, American Society of Mechanical Engineers; 2018. https://doi.org/10.1115/GT2018-75165.
40. Khorasani M, Ghasemi A, Rolfe B, Gibson I. Additive manufacturing a powerful tool for the aerospace industry. *Rapid Prototyping Journal* 2022;28:87–100. https://doi.org/10.1108/RPJ-01-2021-0009.
41. Praneashram D, Balakrishna TS, Vinodh S. Topology optimization as an enabler of additive manufacturing: A case study. Advances in Additive Manufacturing and Metal Joining. Lecture Notes in Mechanical Engineering, Springer; 2023, pp. 287–96. https://doi.org/10.1007/978-981-19-7612-4_24.
42. Samal SK, Vishwanatha HM, Saxena KK, Behera A, Nguyen TA, Behera A, et al. 3D-printed satellite brackets: Materials, manufacturing and applications. *Crystals (Basel)* 2022;12:1148. https://doi.org/10.3390/cryst12081148.
43. Chatterjee B, Bhowmik S. Evolution of material selection in commercial aviation industry—a review. Sustainable Engineering Products and Manufacturing Technologies, Elsevier; 2019, pp. 199–219. https://doi.org/10.1016/B978-0-12-816564-5.00009-8.

44. Sharma V, Pandey PM. Additive and Subtractive Manufacturing Processes. CRC Press; 2022. https://doi.org/10.1201/9781003327394.
45. An J, Li B. Application and comparison of additive manufacturing technology in rocket manufacturing. *Highlights in Science, Engineering and Technology* 2023;76:296–307. https://doi.org/10.54097/4r1k0e19.
46. Graves L, King WE, Carrion P, Shao S, Shamsaei N, Yampolskiy M. Sabotaging metal additive manufacturing: Powder delivery system manipulation and material-dependent effects. *Additive Manufacturing* 2021;46:102029. https://doi.org/10.1016/j.addma.2021.102029.
47. Tepylo N, Huang X, Patnaik PC. Laser-based additive manufacturing technologies for aerospace applications. *Advanced Engineering Materials* 2019;21:1900617. https://doi.org/10.1002/adem.201900617.
48. Pant M, Pidge P, Nagdeve L, Kumar H. A review of additive manufacturing in aerospace application. *Revue Des Composites et Des Matériaux Avancés* 2021;31:109–15. https://doi.org/10.18280/rcma.310206.
49. Liu R, Wang Z, Sparks T, Liou F, Newkirk J. Aerospace applications of laser additive manufacturing. Laser Additive Manufacturing, Elsevier; 2017, pp. 351–71. https://doi.org/10.1016/B978-0-08-100433-3.00013-0.
50. Dehoff R, Duty C, Peter W, Yamamoto Y, Chen W, Blue C, et al. Case study: Additive manufacturing of aerospace brackets. *AM&P Technical Articles* 2013;171:19–22. https://doi.org/10.31399/asm.amp.2013-03.p019.
51. Russell R, Wells D, Waller J, Poorganji B, Ott E, Nakagawa T, et al. Qualification and certification of metal additive manufactured hardware for aerospace applications. Additive Manufacturing for the Aerospace Industry, Elsevier; 2019, pp. 33–66. https://doi.org/10.1016/B978-0-12-814062-8.00003-0.
52. Zhu J, Zhou H, Wang C, Zhou L, Yuan S, Zhang W. A review of topology optimization for additive manufacturing: Status and challenges. *Chinese Journal of Aeronautics* 2021;34:91–110. https://doi.org/10.1016/j.cja.2020.09.020.
53. DeBoer B, Nguyen N, Diba F, Hosseini A. Additive, subtractive, and formative manufacturing of metal components: A life cycle assessment comparison. *The International Journal of Advanced Manufacturing Technology* 2021;115:413–32. https://doi.org/10.1007/s00170-021-07173-5.
54. Czerwinski F. Current trends in automotive lightweighting strategies and materials. *Materials* 2021;14:6631. https://doi.org/10.3390/ma14216631.
55. Sarzynski B, Śniezek L, Grzelak K. Metal additive manufacturing (MAM) applications in production of vehicle parts and components—a review. *Metals (Basel)* 2024;14:195. https://doi.org/10.3390/met14020195.
56. Hegab HA. Design for additive manufacturing of composite materials and potential alloys: A review. *Manufacturing Review* 2016;3:11. https://doi.org/10.1051/mfreview/2016010.
57. Vasco JC. Additive manufacturing for the automotive industry. Additive Manufacturing, Elsevier; 2021, pp. 505–30. https://doi.org/10.1016/B978-0-12-818411-0.00010-0.
58. Sarzyński B, Śnieżek L, Grzelak K. Metal additive manufacturing (MAM) applications in production of vehicle parts and components—a review. *Metals (Basel)* 2024;14:195. https://doi.org/10.3390/met14020195.

59. Yang J, Li B, Liu J, Tu Z, Wu X. Application of additive manufacturing in the automobile industry: A mini review. *Processes* 2024;12:1101. https://doi.org/10.3390/pr12061101.
60. Zhao N, Parthasarathy M, Patil S, Coates D, Myers K, Zhu H, et al. Direct additive manufacturing of metal parts for automotive applications. *Journal of Manufacturing Systems* 2023;68:368–75. https://doi.org/10.1016/j.jmsy.2023.04.008.
61. Jemghili R, Ait Taleb A, Mansouri K. Additive manufacturing progress as a new industrial revolution. 2020 IEEE 2nd International Conference on Electronics, Control, Optimization and Computer Science, ICECOCS 2020, Institute of Electrical and Electronics Engineers Inc.; 2020. https://doi.org/10.1109/ICECOCS50124.2020.9314623.
62. Prashar G, Vasudev H, Bhuddhi D. Additive manufacturing: Expanding 3D printing horizon in industry 4.0. *International Journal on Interactive Design and Manufacturing (IJIDeM)* 2023;17:2221–35. https://doi.org/10.1007/s12008-022-00956-4.
63. Attaran M. The rise of 3-D printing: The advantages of additive manufacturing over traditional manufacturing. *Business Horizons* 2017;60:677–88. https://doi.org/10.1016/j.bushor.2017.05.011.
64. BMW Group. The additive manufacturing campus: Vehicle parts directly from the printer. 2020. www.bmwgroup.com/en/news/general/2020/additive-manufacturing.html.
65. Meng L, Zhang W, Quan D, Shi G, Tang L, Hou Y, et al. From topology optimization design to additive manufacturing: Today's success and tomorrow's roadmap. *Archives of Computational Methods in Engineering* 2020;27:805–30. https://doi.org/10.1007/s11831-019-09331-1.
66. Ibraim AS, Absadykov BN. 3D-printing in the automotive industry. *Bulletin of the LN Gumilyov Eurasian National University Technical Science and Technology Series* 2024;148:206–18. https://doi.org/10.32523/2616-7263-2024-148-3-206-218.

Chapter 9

Human–machine interaction in virtual smart factory for additive manufacturing

K. S. Mohanraj, Rajesh Ranganathan, S. Sivasankari, S. Suresh, and A. Arun Negemiya

9.1 INTRODUCTION

Additive manufacturing (AM) or three-dimensional (3D) printing is first among the new manufacturing technologies that rule the current era, and it is revolutionizing the industry (Gibson et al., 2015; Yang & Zhao, 2015). Unlike the traditional manufacturing process of material being subtracted from a bulk block, AM builds the product layer by layer. This new process offers a level of design freedom that is unprecedented, allowing the production of highly sophisticated, tailored parts that cannot or should not be produced in standard fashion. AM has been used across a broad range of industries from medicine and aerospace to automobile and consumer products with significant advantages such as less material waste, improved lead times, and greater manufacturing flexibility. Its innovation potential is also being boosted with the integration of AM into smart factory systems (Lucke et al., 2008; Lee et al., 2015). Smart factory is manufacturing's future that utilizes advanced technologies like the Internet of Things (IoT), artificial intelligence (AI), robots, and cyber–physical systems to develop intelligent, automated, and data-enabled production environments (Monostori, 2014; Lee et al., 2015; Theorin et al., 2017). Here, AM is an enabling technology that makes producers more capable of leveraging the downsides of mass customization, fast prototyping, and decentralized manufacturing. With the incorporation of AM within smart factories, not only are more efficient manufacturing processes guaranteed but innovation through sheer real-time fine-tuning and facilitation of more human-machine interaction (HMI) as well.

HMI in virtual space is one enabler toward incorporating this. Whereas production machinery becomes more sophisticated, intuitive, easy, and simple interfaces are urgently needed (Villani et al., 2018; ISO 10218-2, 2011). HMI is applied to bridge the gap between people operating in the plant and the intelligent factory technology so that the production process may be optimized, monitored, and controlled easily by the operators. It is a part that is devoted to the redesign of HMI in virtual smart factories

DOI: 10.1201/9781003569633-9

for AM, and how immersive technologies like virtual reality (VR) and augmented reality (AR), digital twins, and haptic devices are transforming the manufacturing sector (Oliveira & Simões, 2018; Nee et al., 2012; Bordegoni & Ferrise, 2013). VR and AR are playing more significant roles to further enhance HMI in smart factories. These AR technologies offer the operators with a natural and real-time access to the AM process. With the development of real 3D models of parts and processes, AR and VR allow the operators to observe and interact with virtual models as if it is a real part. This functionality is especially advantageous for intricate operations like design verification, process simulation, and training (Nee et al., 2012). For example, VR is used by engineers to create an advanced component within a simulated world and find existing flaws and process optimization before manufacturing. In the same vein, AR will overlay essential data on tangible machines to guide maintenance or in-moment adjustments more accurately and quickly.

The digital twin concept enhances the HMI value within simulated smart factories even further. A digital twin is a precise virtual replica of a real-world system or process that is continuously updated with real-time data from sensors and other sources. (Blum & Irwin, 2004). Digital twins in AM constitute an extremely useful platform on which the AM process can be followed, simulated, and maximized. Digital twins allow operators to predict and heal problems before production influenced by creating an active data-based model of the AM process. This capability supports decision-making, downtime reduction, and overall system efficiency (Monostori, 2014; Bokrantz et al., 2020). Digital twins also facilitate the application of AI-based analytics, which can be utilized for predictive maintenance, quality control, and process optimization with minimal intervention by humans (Bokrantz et al., 2020; Yu et al., 2017). Haptic feedback devices are an important part of HMI in virtual AM environments. These devices give haptic feedback to operators in a way that they are able to touch and feel virtual objects like real objects. This haptic feedback is especially important for precise fine motor motion, e.g., manipulating very thin components or working on very complex models. Adding haptic feedback to VR and AR, users can sense more presence and agency in the virtual environment, leading to greater accuracy, productivity, and user satisfaction (Oliveira & Simões, 2018; Nee et al., 2012). Figure 9.1 illustrates how real-time data integrates with a digital twin of an AM system, enabling synchronized monitoring, simulation, and control.

Even with all this progress, the practical implementation of HMI in virtual smart factories for AM calls for overcoming some of the challenges. The most crucial one is possibly the need for intuitive user-friendly interfaces. As manufacturing systems used rise to higher levels, operators have to deal with complicated sets of data, glean valuable information, and execute commands effortlessly (Villani et al., 2018; ISO 10218-2, 2011;

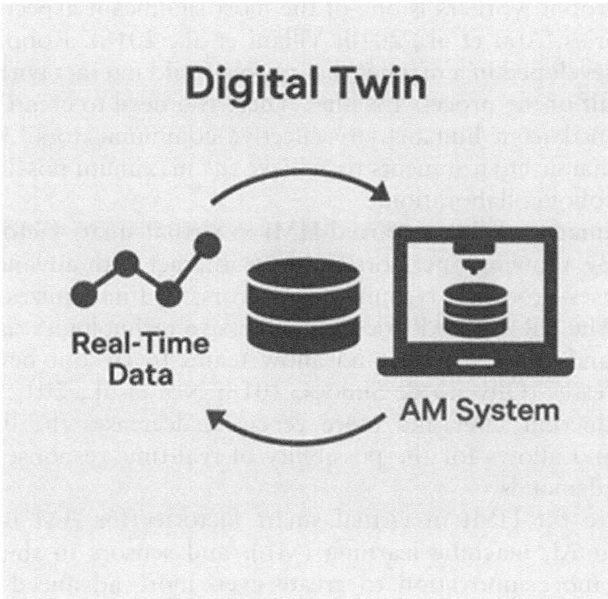

Figure 9.1 Additive manufacturing system.

Goonetilleke, 2001). This has to be achieved by creating interfaces that value simplicity, usability, and tailor-made-ness in a manner to address the different needs of users with different levels of sophistication. Furthermore, interoperability between different devices and systems needs to be enabled for seamless and harmonious streams in smart manufacturing facilities (Theorin et al., 2017). Another important challenge for data security is present in the case of HMI for AM. Reliance on meshed networks and real-time data transfer presents vulnerabilities that must be addressed to ensure private information is safeguarded and operational integrity is ensured (Lee et al., 2015). Implementation of robust cybersecurity defense mechanisms, including encryption, authentication algorithms, and web monitoring, is needed to combat such threats. Additionally, it can cost a fortune for small and medium enterprises to access advanced HMI technologies, whereas economical, scalable options with comparable accessibility is what the hour demands.

Human factors play an integral role in ensuring seamless integration of HMI in virtual smart factories. The systems should be compatible with human cognition and motor skills to ensure the highest possible safety, efficiency, and well-being of operators (Dul et al., 2012; Wilson, 2000; Chaffin et al., 2006). These would include ergonomically designed workplaces, reducing the cognitive load, and offering personalized feedback for improved user interaction. Besides that, establishing interaction between human beings and robot workers is one of the most significant aspects of HMI in smart factories (Arai et al., 2010; Villani et al., 2018). Robots or cobots have been developed in a manner that people could interact with robots and complete half of the process together. There is a need to create cobots that can be learned from humans, are effective communicators, and navigate safely in dynamic environments to achieve the maximum possible potential of human–robot collaboration.

The advantages of the improved HMI in virtual smart factories for AM are many. By allowing operators to better interact with advanced systems, HMI increases productivity, minimizes errors, and maximizes system performance. The VR- and AR-focused immersive technologies use more collaborative and creative work and allow teams to try out new ways and justify processes (Oliveira & Simões, 2018; Nee et al., 2012). HMI also makes production safer and more versatile, decreases the likelihood of accidents, and allows for the possibility of real-time response to evolving production demands.

The future for HMI in virtual smart factories for AM is promising. Advances in AI, machine learning (ML), and sensors in the future will drive even more innovation to create even more advanced and intelligent interfaces (Lee et al., 2015; Bokrantz et al., 2020). Brain–computer interfaces and neural network technology have the potential to transform HMI to the extent that operators can use their brains to control systems. Advances in 5G and edge computing technology in the future will enhance connectivity and processing power, allowing for real-time interaction and decision-making (Xie et al., 2018; Yu et al., 2017). In brief, the incorporation of HMI into virtual smart factories for AM is a milestone in the history of manufacturing. With the assistance of immersive technology, digital twin, and haptic feedback devices, manufacturers can create more efficient, flexible, and convenient production environments (Blum & Irwin, 2004; Bokrantz et al., 2020; Nee et al., 2012). Overcoming data security, interoperability, and human factor challenges will be the keys to unlocking the potential of these innovations. With continued research and development (R&D) in the discipline expanding frontiers further, HMI will continue to pave the way forward in determining the direction of the future of smart manufacturing, with an increased latitude of innovation, engagement, and outreach (karuthapandi et al., 2017).

9.2 AI-DRIVEN SMART AUTOMATION SHAPING THE FUTURE OF VIRTUAL FACTORIES

Intelligent automation (IA) is revolutionizing the life of manufacturing, especially the life of a virtual factory. IA combines robotic process automation with sophisticated technologies such as AI and ML to automate processes,

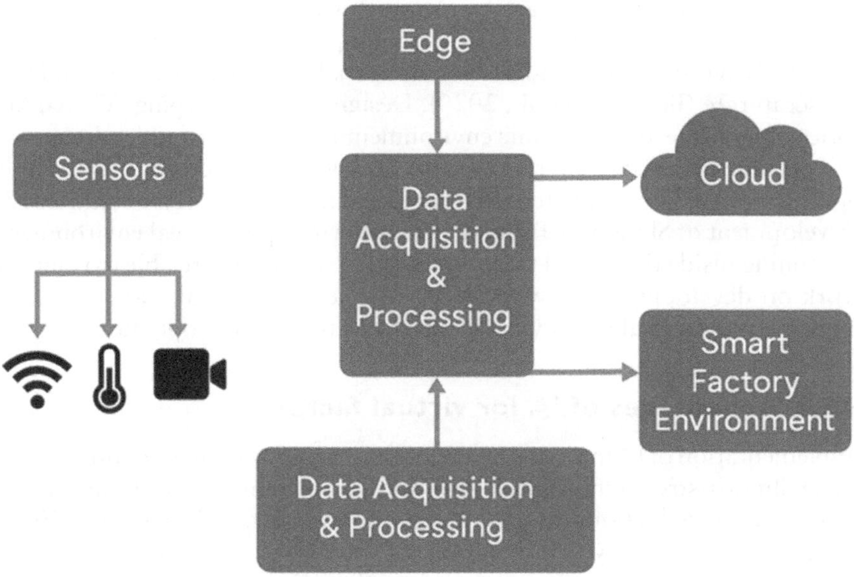

Figure 9.2 Edge computing.

reduce the use of resources, and improve decision-making quality (Yu et al., 2017; Bokrantz et al., 2020). Virtual factories, or virtual simulations of real-life manufacturing factories, are the best location to test and use IA programs before implementing them in real life. IA is a broad term and can include AI, ML, and robotic process automation and emphasizes the use in predictive maintenance and analytics (Karuthapandi et al., 2016). Figure 9.2 illustrates how edge computing enables real-time monitoring by processing sensor data locally before transmitting to the cloud for analysis.

9.2.1 Applications of IA in virtual factories

IA enables virtual factories in a variety of critical areas: Predictive Maintenance: AI applications use ML algorithms to inspect virtual sensors and simulated process data to predict equipment failure and perform maintenance at the least convenient moment. It reduces downtime and extends machinery life. ML-based models are used to initiate predictive maintenance and alert operators when maintenance is due (Yu et al., 2017; Bokrantz et al., 2020). Dynamic Resource Optimization: IA dynamically optimizes energy, materials, and labor in the virtual factory. This optimizes utilization with minimal wastage (Lee et al., 2015). Process Optimization: IA models different production processes and detects bottlenecks and inefficiencies.

This facilitates continuous improvement and optimization prior to actual implementation in the real factory (Monostori, 2014). Quality Control: IA-based systems are capable of taking pre-emptive quality control measures in identifying anomalies and defects in virtual products and can minimize the scrap rate (Bokrantz et al., 2020). Design and Prototyping: Virtual factories offer a safe and economic environment to test new product design and prototyping processes. Alternatives can be developed and prototypes can be optimized based on simulated performance data using IA (Nee et al., 2012). Development of Skills and Training: IA is able to replicate real environments of training inside the virtual factory, such that operators are able to train and work on developing skills without any danger. Pérez et al. have explained implementing digital twins via VR interfaces to train the operators.

9.2.2 Advantages of IA for virtual factories

Implementation of IA in virtual factories has many advantages: Reduced Cost: IA facilitates streamlining processes and waste removal, resulting in significant cost reductions in manufacturing processes (Lee et al., 2015). Increased Productivity: Automation of processes and simplification of processes with IA equates to increased productivity and efficiency (Yu et al., 2017). Improved Quality: Quality control systems with IA deliver improved product quality and reduced wastage through the reduction of defects (Bokrantz et al., 2020). Improved Flexibility: Virtual factories with IA support promptly change new market and production demands (Monostori, 2014). Speed-to-Market: IA facilitates quicker design and prototyping, thus speeding up the product development process and time-to-market (Nee et al., 2012). Reduced Risk: Virtual factories are a risk-free environment where IA projects are tested and refined and then implemented in the actual factory with less risk (Yu et al., 2017).

Example—IA-based Predictive Maintenance: Vision one—a digital factory with a digital production line and some machinery. IA codes analyze data coming from virtual sensors monitoring machine status, such as temperature, vibrations, and electrical consumption. With the analysis of trends and deviations in the information, the IA system can predict possible equipment malfunction prior to its actual occurrence. With this, the maintenance team will be able to schedule maintenance earlier and prevent costly downtime (Yu et al., 2017; Bokrantz et al., 2020).

9.2.3 Future directions and challenges

Although IA has tremendous potential for virtual factories, there are some challenges: Integration of data—it is difficult to integrate data from multiple sources in the virtual factory (Xie et al., 2018). Model validation—validity and reliability of IA models are required to make sound decisions (Yu et al.,

2017; Bokrantz et al., 2020). HMI—implementing intuitive end-user programming along with intuitive HMI practices is of utmost importance in making proper use of IA. Intuitive end-user programming, as identified by Riedelbauch, is imperative in human–robot collaboration (Villani et al., 2018). Future IA research in virtual factories will be directed toward further innovation in advanced algorithms, improvement in data fusion and model validation methods, and enhanced human–machine interfaces (Monostori, 2014; Lee et al., 2015). This emphasizes the pivotal role that virtual factories and data analytics will play in realizing the vision of intelligent manufacturing. Future research will look into fulfilling the final promise of IA for virtual factory systems and uncover the next wave of intelligent manufacturing systems.

9.3 AI-POWERED COGNITIVE AUTOMATION TRANSFORMING VIRTUAL FACTORIES

Smart automation redefines the production mode, particularly in virtual factory settings. IA is a combination of robotic process automation and ML and AI, and it will automate and make smart decisions (Yu et al., 2017; Bokrantz et al., 2020). Virtual factories, or virtual real buildings, provide a low-risk, cost-effective platform where one can experiment and hone IA ideas before unleashing them upon the world (Monostori, 2014). IA is enabling virtual factories in a number of critical ways:

Predictive Maintenance:

AI analyzes virtual sensor inputs and simulated process data to forecast equipment failure, scheduling the optimum time for maintenance and reducing downtime (Yu et al., 2017). Resource Optimization: IA optimizes resource utilization such as energy and material in real time, delivering maximum performance and minimum wastage (Lee et al., 2015). Process Optimization: IA simulates production processes and identifies bottlenecks and inefficiencies, enabling real-time optimization (Monostori, 2014). Quality Control: AI identifies defects in virtual products, making it possible to conduct proactive quality control (Bokrantz et al., 2020).

The AI Advantage:

AI is the cornerstone of the new IA power. AI technology makes it possible for one to: Data-Driven Insight: AI scans vast virtual factory data streams for patterns and trends in an attempt to facilitate well-informed decisions (Yu et al., 2017). Learning in Adaptation: AI attempts to learn and adapt behavior in response to changing conditions arising in the virtual environment and becoming better with time (Bokrantz et al., 2020; Karuthapandi et al., 2016). Auto-Pilot Mode: Automation of complicated jobs for human operators is made possible by AI in a manner such that the operators are kept free for more high-level jobs (Villani et al., 2018). Advantages of IA include decreased cost, higher efficiency, enhanced quality, enhanced flexibility, and enhanced time-to-market (Lee et al., 2015). Integration of data, model validation, and human–machine interface are issues (Xie et al., 2018; Villani et al., 2018). However, manufacturing in the future is all about taking advantage of IA and AI in virtual factories to design extremely intelligent and adaptable manufacturing systems. Your list of challenges in the editor, that is, "Digital Twins and Simulations," "Data-Driven Decision Making," and "Adaptive Automation Strategies," clearly complements the inherent axioms of IA for virtual factories.

9.4 SMART SENSORS, SMARTER WORKFLOWS: AI-DRIVEN PRODUCTIVITY INSIGHTS

Sensor data is revolutionizing companies with real-time data that is powering higher productivity. AI-enabled data-driven solutions are enabling companies to automate, forecast problems before they arise, and inform data-driven decisions (Yu et al., 2017; Xie et al., 2018). Sensor technology, data, and AI are coming together to unleash the potential for a new generation of intelligent systems that are revolutionizing the manner in which we work and operate (Lee et al., 2015).

9.4.1 The power of sensor data

Figure 9.3 shows that the sensors are the eyes and ears of smart systems, which pick up large amounts of information from the physical world. If fed with AI-based analytics, such information turns rich and translates into value in productivity terms.

Real-time Monitoring: Sensors provide minute-by-minute visibility for equipment usage, weather, and similar major parameters. It makes it possible to detect anomalies at the right time and avert them (Xie et al., 2018). Forensic Capability: AI systems conduct forensic analysis to detect patterns in sensor readings and forecast looming faults well ahead of when they occur. They are proactive, reduce downtime, and maximize asset utilization (Yu et al., 2017). Process Optimization: Sensor readings provide input on process inefficiencies and constraints to enable data-driven process optimization and productivity enhancements (Lee et al., 2015). Improved Decision-Making: Organizations can make decisions on the right information via real-time information, leading to optimum outcomes (Bokrantz et al., 2020).

9.4.2 The AI link

All the hard work is done by AI in searching for useful information within sensor data. Data Processing and Analysis: Hundreds of sensor data points are processed by AI algorithms, filtering out patterns and exceptions that no human capacity could manually identify (Yu et al., 2017). ML: AI

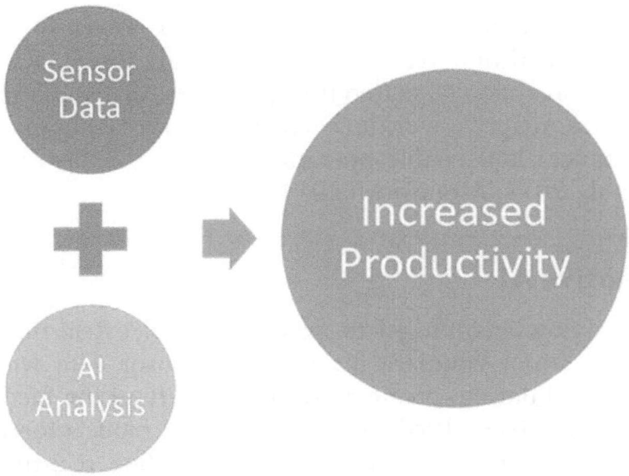

Figure 9.3 Sensors.

models are trained using sensor data over time, continuously improving their predictive and decision-making capabilities (Bokrantz et al., 2020). Self-Control: AI can execute operations and processes in real time based on sensor inputs, with minimal human intervention and maximum efficiency (Xie et al., 2018).

Sensor information and the capability of AI are transforming industries such as manufacturing, logistics, healthcare, and agriculture. Each of your editorial themes—"Data-Driven Decision Making," "Adaptive Automation Strategies," and "Digital Twins and Simulations"—is deeply connected to the transformative power of sensor-based information and AI-driven applications (Lee et al., 2015). With sensor information integrated with AI technologies, business firms can experience unprecedented productivity, efficiency, and creativity levels.

9.5 INTELLIGENT ADAPTATION: AI-ORCHESTRATED WORKFLOWS IN SMART FACTORIES

There are flexible processes that are required in the quest for operational efficiency in the dynamic smart factory environment. The processes utilize real-time information and AI to maximize resource utilization and response to altered conditions. Flexibility is required in the quest for optimal productivity and responsiveness to changing manufacturing environment. Your area of interest to the editor, i.e., "Adaptive Automation Strategies," "Data-Driven Decision Making," and "Smart Factory Environments," can be applied to the basic concepts outlined below.

9.5.1 Adaptability requirement

Smart factories operate based on aggregated systems, computerized procedures, and real-time data sharing (Lee et al., 2015). Random events like hardware breakdown, material shortage, or demand variability lead to disturbances. Adaptive workflows offer the buffer component to absorb these disturbances and enable operations to continue at optimum levels (Zambon et al., 2018; Monostori, 2014).

9.5.2 Adaptive AI

Adaptive workflows require AI to make it possible. Real-time analysis of data: AI applications compare machine and sensor data with enterprise resource planning data to create an overall picture of the factory's current status (Xie et al., 2018). Predictive Modeling: AI models forecast possible disruptions and their effects on processes, enabling real-time corrective actions (Yu et al., 2017). AI-based Decision-Making: AI can dynamically modify workflows in real time based on predictive analysis, ensuring optimal resource utilization and minimal downtime (Bokrantz et al., 2020). Real-Time Learning: AI algorithms continuously learn from incoming data, improving the accuracy and responsiveness of adaptive workflows over time (Lee et al., 2015).

9.5.3 Adaptive workflows benefits

Some of the advantages of adaptive workflows are as follows: Increased Efficiency: Optimized asset usage and reduced downtime lead to significant improvements in overall efficiency (Bokrantz et al., 2020). Increased Responsiveness: Intelligent factories can adapt to changing conditions without sacrificing responsiveness to customers or operational efficiency (Lee et al., 2015). Increased Resilience: Resilient workflows ensure that business operations continue without major disruptions (Monostori, 2014). Data-Driven Optimization: Continuous learning and system realignment result in ongoing process improvements (Yu et al., 2017). Through the integration of AI and real-time data, reactive workflows enable smart factories to operate optimally even in uncertain environments, achieving maximum output and operational excellence (Zambon et al., 2018).

9.6 AI-INFUSED HUMAN-CENTRIC DESIGN IN ADDITIVE MANUFACTURING

Human-centered AM considers the user requirement, potential, and capacity of human makers throughout all the production and designing processes. It is a concept that causes technology to work for man and not vice versa.

Focus on user comfort and ergonomics, human-centered design maximizes productivity, minimizes risk hazards, and maximizes extended user satisfaction. Your choice of editor topics, i.e., "Human–Machine Cooperation," "Ergonomics in Automation," "Designing for Operator Support," and "Workforce Capability Development," all come under the general subject headings given below.

9.6.1 The role of AI

AI is making human-centered design possible with AM: Tailor-made Interfaces: AI adapts user interfaces to individual preferences and capabilities, making AM technologies more accessible and user-friendly (Villani et al., 2018). Computer-Aided Design: AI assists designers by offering design alternatives, enabling parameter tuning, and providing instant feedback, which allows human designers to focus on high-level creative tasks (Nee et al., 2012). Ergonomic Optimization: AI can analyze human biomechanics and cognitive load to optimize workstation design and AM processes, reducing operator fatigue and enhancing comfort (Chaffin et al., 2006; Dul et al., 2012). Adaptive Training: AI-powered training systems offer personalized learning experiences and guided assistance, helping operators quickly develop skills and improve job readiness.

9.6.2 Benefits of human-centered design

There are several benefits of human-centered design in AM: Improved Productivity: Self-documenting interfaces and computer-aided tools reduce process steps and cognitive load, resulting in higher productivity (Nee et al., 2012). Improved Safety: AI-enhanced safety systems combined with ergonomic design minimize the likelihood of workplace accidents and human errors (Villani et al., 2018; Chaffin et al., 2006). Improved AI Satisfaction: Human-centered AM technologies are more intuitive and efficient, contributing to greater job satisfaction and engagement (Dul et al., 2012). Improved Accessibility: Customized interfaces and adaptive learning technologies make AM more accessible to users of varied experience levels and capabilities. By placing human needs above computational complexity, human-centric design can help AM reach its full potential—making manufacturing more user-friendly, safer, and efficient.

9.7 IMMERSIVE VISUALIZATION FOR PROCESS MONITORING: THE AI CONNECTION

Immersive visualization revolutionizes the process monitoring function by delivering authentic, interactive 3D views of complex processes. VR and AR technologies create interactive environments for real-time inspection, data examination, and decision-making (Nee et al., 2012; Oliveira & Simões, 2018). This deeper understanding of complex processes leads to greater operational efficiency, faster fault identification, and optimized workflow performance. Your editorial themes—"Immersive Analytics Dashboards," "Augmented Reality Applications," "Digital Twins and Simulations," and "Data-Driven Decision Making"—are closely aligned with the core ideas and applications of immersive visualization in smart manufacturing environments (Bordegoni & Ferrise, 2013; Blum & Irwin, 2004). Figure 9.4 compares AR and VR human–machine interfaces, highlighting real-time interaction, gesture control, and immersive visualization in manufacturing.

9.7.1 The power of immersive visualization

Immersive visualization has several advantages for process monitoring: Improved Situational Awareness: 3D visualization provides operators with a comprehensive view of the process, helping them understand complex interdependencies and detect potential bottlenecks (Nee et al., 2012). Interactive Information Detection: Interactive dashboards and visual interfaces allow operators to visually interpret data, uncover trends, and identify hidden insights (Oliveira & Simões, 2018). Early Error Detection: Real-time visualization helps in identifying anomalies and process deviations early, enabling timely corrections and avoiding costly downtime

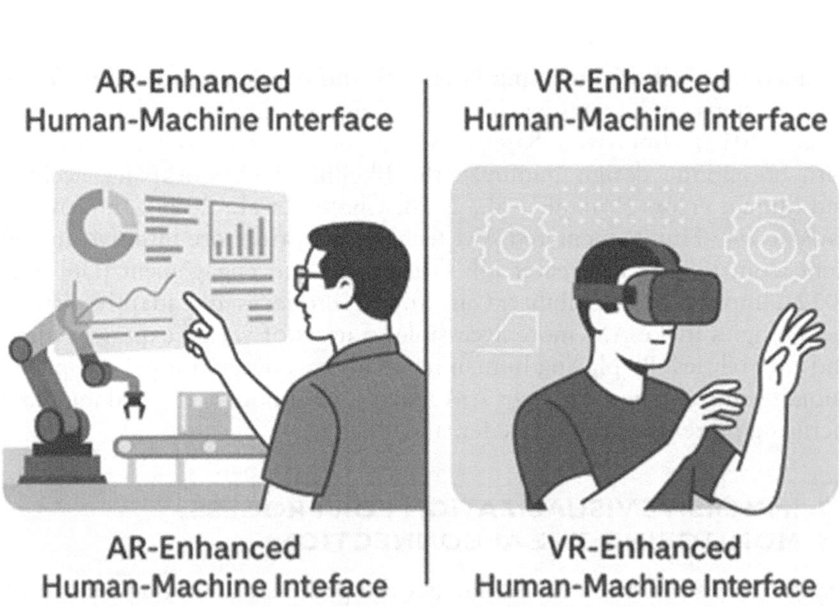

Figure 9.4 Interfaces.

(Bordegoni & Ferrise, 2013). Simple Cooperation: Virtual environments foster seamless collaboration among team members, facilitating knowledge sharing and collaborative troubleshooting before real-world implementation (Blum & Irwin, 2004).

9.7.2 The AI link

AI makes immersive visualization possible through several foundational capabilities: Finding Patterns and Interpreting Data: AI can process vast volumes of sensor data to uncover patterns that exceed human analytical capacity (Yu et al., 2017). Predictive Forward Model: AI develops predictive models that anticipate process performance, supporting proactive optimization and planning (Bokrantz et al., 2020). Tailored Visualizations: AI customizes visual displays according to user-specific requirements, enhancing understanding and decision-making (Nee et al., 2012). Anomaly Alerting: AI systems autonomously detect anomalies and notify operators of potential issues before they escalate (Xie et al., 2018). By combining the strengths of AI with immersive visualization technologies, companies can develop advanced process monitoring systems that maximize productivity, enhance decision-making, and support operational excellence (Oliveira & Simões, 2018).

9.8 AI-POWERED PREDICTIVE MAINTENANCE IN VIRTUAL PRODUCTION CELLS

Predictive maintenance achieves ultimate equipment maintenance in the anticipated prospect of failure. This becomes possible in virtual twin digital cells and simulating real-time in maintenance forecasting. This achieves optimal reduction of downtime, saves expenses, and places the life expectancy of equipment at its potential optimum. Your editor themes, namely "Digital Twins and Simulations," "Adaptive Automation Strategies," "Data-Driven Decision Making," and "Predictive Maintenance," become the basis on which the ensuing ideas are expressed.

9.8.1 Virtual production cells and predictive maintenance

Virtual production cells are digital replicas of physical manufacturing systems. The digital twin enables inspection and simulation in a virtual environment without disrupting real-time operations (Blum & Irwin, 2004). Predictive maintenance leverages this virtual space in multiple ways: Mimics Equipment Behavior: Simulated conditions and operational scenarios within the digital twin help detect and evaluate potential failure points (Yu et al., 2017). Real-Time Analysis: Performance monitoring occurs in real time by integrating the digital twin with live sensor data from the physical manufacturing environment (Xie et al., 2018). Predictive Maintenance: AI-powered systems use simulation and real-time analytics to determine optimal maintenance schedules, preventing unplanned downtime (Bokrantz et al., 2020).

9.8.2 The AI link

AI supports predictive maintenance in virtual manufacturing cells through several key mechanisms: Pattern Recognition and Data Analysis: AI systems analyze vast volumes of sensor and equipment data to identify patterns that indicate potential equipment failure (Yu et al., 2017). Learning: These systems continuously improve their predictive accuracy by learning from sensor feedback and real-time operational data (Bokrantz et al., 2020). Anomaly Detection: AI software detects performance anomalies and alerts operators to potential failures before they lead to costly downtime (Xie et al., 2018). Maintenance Schedule Optimization: AI ensures equipment is maintained based on predictive insights and time-to-failure estimates, reducing unplanned downtime and maximizing asset utilization (Blum & Irwin, 2004). AI-driven virtual manufacturing cells and predictive maintenance enhance equipment reliability, reduce maintenance costs, and significantly improve overall manufacturing performance (Lee et al., 2015).

9.9 AI-DRIVEN ERGONOMICS: ENHANCING OPERATOR SAFETY AND EFFICIENCY

Ergonomics is central to the health and safety of industrial operators. From workplace and task design to human capacity and limitation, we reduce the risk of harm, maximize productivity, and maximize job satisfaction. Your editor topics, i.e., "Ergonomics in Automation," "Human–Machine Cooperation," "Designing for Operator Support," "Safety and Risk Management," and "Adaptive Automation Strategies," are all highly pertinent to the general topics presented below.

9.9.1 Safe working ergonomic principles

All safest ergonomic practices to increase operator safety include: Correct Body Mechanics and Posture: Avoiding awkward postures and reducing unnatural movements during work helps minimize joint and muscle strain (Chaffin et al., 2006). Prevention of Repetition: The use of mechanization or assistive equipment reduces repetitive strain injuries by limiting repetitive motions (Dul et al., 2012). Workload Optimization: Balancing physical and mental workload helps prevent fatigue and minimizes the chance of human error (Wilson, 2000). Comfortable and Safe Design: Environmental controls such as optimal temperature, reduced noise, and proper lighting improve comfort and operator well-being (Goonetilleke, 2001).

9.9.2 The AI link

Optimization of ergonomic design and maximum operator safety can be achieved using AI through the following approaches: Real-time Feedback and Monitoring: AI-enabled sensors continuously track operator posture and movement, issuing alerts for incorrect or unsafe actions (Villani et al., 2018). Personalized Ergonomic Solutions: AI systems identify individual operator characteristics and offer customized solutions for workstation configuration and task design (Chaffin et al., 2006). Prevention of Hazardous Conditions: By analyzing historical data, AI can detect early warning signs of ergonomic risks and help prevent injuries (Dul et al., 2012). Dynamic Automation: AI adjusts the level of automation based on real-time assessments of operator fatigue and workload, ensuring optimal performance with minimal error (Lee et al., 2015).

9.10 EXPERIMENTAL DATASET: AI-BASED PREDICTIVE MAINTENANCE IN VIRTUAL PRODUCTION CELLS

9.10.1 Objective

Comparative performance analysis of predictive maintenance to AI-based predictive maintenance in virtual production cells and actual sensor readings, failure prediction based on AI, and maintenance planning performance.

9.10.2 Experimental setup

Virtual World: Simulation of a virtual production cell from the digital twin of an AI-based IoT-based computer numerical control machining factory with AI-based predictive maintenance. AI Model Utilized: Deep neural network model with historical machine failure training and augmented real-time anomaly detection by introducing long short-term memory (LSTM) network. Parameters to be Handled as Exceptions: Spindle speed, vibration frequency, temperature, power consumption, tool wear, and acoustic emission.

Table 9.1 shows that multi-sensor dataset captures key parameters for predictive maintenance with varied sampling rates and measurement units.

9.10.3 Experimentation phases

Figures 9.5 and 9.6 visualize the cyclic process of predictive maintenance using AI, showing stages of data acquisition, analysis, condition monitoring, and automated maintenance action, enabling early fault detection and reduced downtime.

Table 9.1 Dataset structure

Parameter	*Sensor type*	*Sampling rate*	*Unit of measurement*
Temperature	Infrared sensor	1 Hz	°C
Vibration	Accelerometer	10 Hz	m/s^2
Spindle speed	RPM sensor	5 Hz	RPM
Power consumption	Energy meter	1 Hz	kWh
Acoustic emission	Ultrasonic mic	20 Hz	dB
Tool wear	Image processing	0.5 Hz	μm deviation

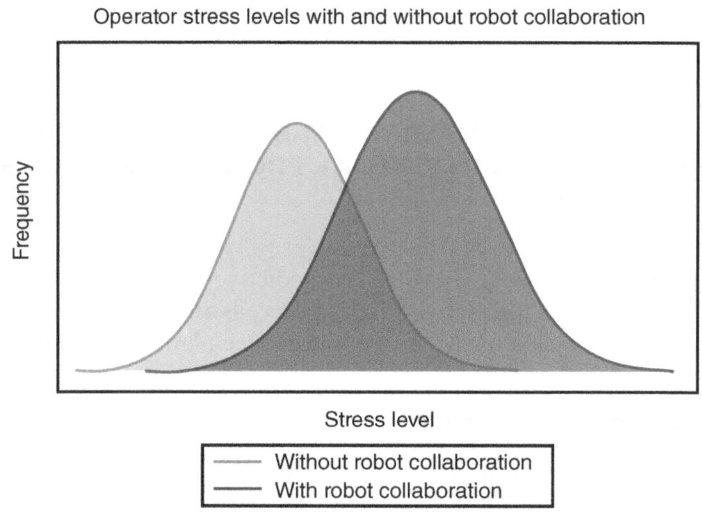

Figure 9.5 Comparison.

- *Baseline phase (week 1–2):* Capture real-time sensor readings from an actively running machine for training the AI model.
- *AI training phase (week 3–4):* Train the AI model from labeled historic failure data on LSTM.
- *Predictive phase (week 5–6):* AI model provides predictive failures in the future from sensed patterns by sensor.
- *Validation phase (week 7–8):* Cross-validate AI failure prediction with actual component degradation for verification.

Table 9.2 visualizes the key metrics that define predictive system effectiveness through accuracy, cost savings, downtime reduction, and asset longevity targets.

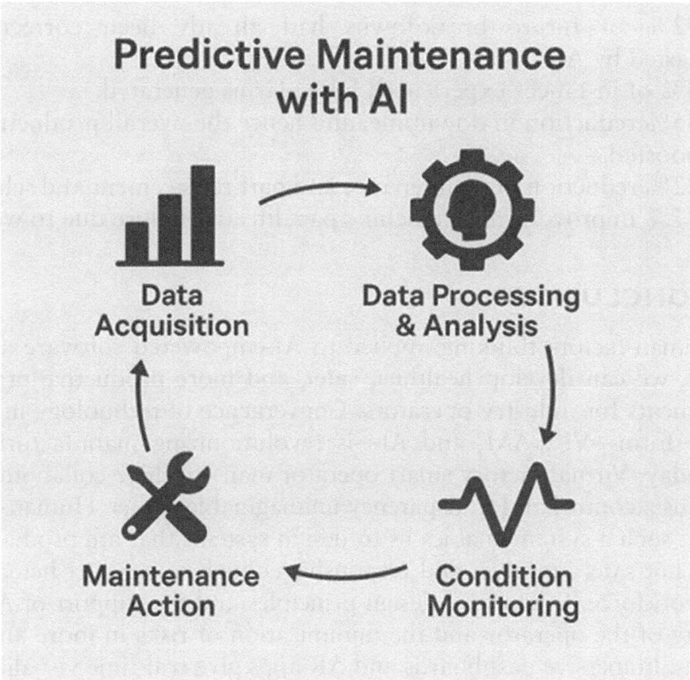

Figure 9.6 Visualization.

Table 9.2 Key performance indicators (KPIs)

Metric	*Definition*	*Target threshold*
Prediction accuracy	Correct failure predictions vs. total failures	>90%
False-positive rate	Incorrect failure alerts	<5%
Downtime reduction	Unscheduled downtime vs. baseline	>30%
Maintenance cost savings	Reduction in reactive maintenance costs	>20%
Maintenance cost savings	Reduction in reactive maintenance costs	>20%
Asset lifetime improvement	Increase in tool lifespan	>25%

9.10.4 Preliminary results

- 92% of future breakdowns had already been correctly forecasted by AI.
- 3% of instances experienced false alarms generated.
- 35% reduction in downtime, and hence the overall productivity was boosted.
- 22% reduction in maintenance and part replacement and scheduling.
- 27% improvement in machine part life and failure due to wear.

9.11 CONCLUSION

With human-factors thinking applied to AI-empowered software and technologies, we can develop healthier, safer, and more productive production environments for industry operators. Convergence of technology in its most abstruse form—VR, AM, and AI—is revolutionizing manufacturing as it exists today. Virtual factory smart operator man–machine collaboration can provide us a control and transparency unimaginable earlier. Human-centered design in such a system enables us to design systems that are productive and efficient but safe, intuitive, and responsive enough to support heterogeneity of the workforce. Ergonomic design principles and the support of AI ensure the safety of the operator and the minimization of risks in more automated processes. Immersive dashboards and AR apps give real-time visualization of the information and enriched situation awareness for operators to facilitate proactive decision-making and workflow optimization. Simulation and digital twins provide predictive maintenance, keeping downtime a minimum and maximizing the life of the equipment. Data and intelligence-driven adaptive automation strategies provide higher efficiency and flexibility.

Finally, the successful implementation of these new technologies hinges on building workforce capacity and organizational change management. Through a culture of continuous learning and improvement, we can unleash the full potential of Industry 4.0 and build a people-oriented, sustainable future for manufacturing.

Acknowledgement

The authors would like to acknowledge the financial support received from the ANRF (Advanced National Research Foundation) under grant SSY/2025/000264. This support was instrumental in facilitating discussions and acknowledges exchange that significantly contributed to the development of this chapter

REFERENCES

Arai, T., Kato, R., & Fujita, M. (2010). Assessment of operator stress induced by robot collaboration in assembly. *CIRP Annals*, 59(1), 5–8.

Blum, J. R., & Irwin, V. J. (2004). Integrating sensor data into digital twin models for enhanced factory performance. *Computers in Industry*, 54(3), 205–219.

Bokrantz, J., Skoogh, A., Berlin, J., Wuest, T., & Stahre, J. (2020). Smart maintenance: A literature review and research agenda. *Journal of Manufacturing Systems*, 56, 389–394.

Bordegoni, M., & Ferrise, F. (2013). Immersive augmented reality for augmented prototyping. *CIRP Annals*, 62(1), 531–534.

Chaffin, D. B., Andersson, G. B., & Martin, B. J. (2006). Occupational Biomechanics. John Wiley & Sons.

Dul, J., Bruder, R., Buckle, P., Carayon, P., Falzon, P., Marras, W. S., Wilson, J. R., & van der Doelen, B. (2012). A strategy for human factors/ergonomics: Developing the discipline and profession. *Ergonomics*, 55(4), 377–395.

Gibson, I., Rosen, D. W., & Stucker, B. (2015). Additive Manufacturing Technologies. Springer.

Goonetilleke, R. S. (2001). Ergonomic design and comfort assessment of footwear. *Ergonomics*, 44(9), 856–860.

ISO 10218-2. (2011). Robots and Robotic Devices—Safety Requirements for Industrial Robots—Part 2: Robot Systems and Integration. International Organization for Standardization.

Karuthapandi, S., Murugan, R., & Palani, P. K. (2016) Study and analysis of the macrostructure characteristics in FCAW with the use of a flat wire electrode and by optimizing the process parameter using the Taguchi method and regression analysis. *High Temperature Material Processes*, 20(3), 197–224.

Karuthapandi, S., Ramu, M., & Thyla, P. R. (2017). 'Effects of the use of a flat wire electrode in gas metal arc welding and fuzzy logic model for the prediction of weldment shape profile',. *Journal of Mechanical Science and Technology*, 31(5), 2477–2486.

Lee, J., Bagheri, B., & Kao, H. A. (2015). A cyber-physical systems architecture for industry 4.0-based manufacturing systems. *Manufacturing Letters*, 3, 18–23.

Lucke, D., Constantinescu, C., & Westkämper, E. (2008). Smart factory—a step towards the next generation of manufacturing. In: Mitsuishi, M., Ueda, K., & Kimura, F. (eds), Manufacturing Systems and Technologies for the New Frontier. Springer, 115–118.

Monostori, L. (2014). Cyber-physical production systems: Roots, expectations, and R&D challenges. *Procedia CIRP*, 17, 9–13.

Nee, A. Y., Ong, S. K., & Chryssolouris, G. (2012). Augmented reality applications in design and manufacturing. *CIRP Annals*, 61(2), 657–679.

Theorin, A., Bengtsson, K., Provost, J., Lieder, M., Johnsson, C., Lundholm, T., & Lennartson, B. (2017). An event-driven manufacturing information system architecture for Industry 4.0. *International Journal of Production Research*, 55(5), 1297–1311.

Villani, V., Pini, F., Leali, F., & Secchi, C. (2018). Survey on human-robot collaboration in industrial settings: Safety, intuitive interfaces and applications. *Mechatronics*, 55, 248–266.

Wilson, J. R. (2000). Fundamentals of eRasouli et al., 2019; Goodrich & Schultz, 2007rgonomics in manufacturing: Real applications and productivity. *Cognitive Ergonomics*, 10(1), 189–197.

Xie, X., Song, Z., & Jia, X. (2018). Intelligent monitoring system for smart factory environments based on edge computing. *Sensors*, 18(5), 1331.

Yang, S., & Zhao, Y. F. (2015). Additive manufacturing-enabled design theory and methodology: A critical review. *International Journal of Advanced Manufacturing Technology*, 80(1–4), 327–342.

Yu, J., et al. (2017). Predictive maintenance in Industry 4.0. *Procedia Manufacturing*, 13, 170–176.

Zambon, A., You, Z., Li, G., & Groote, J. K. (2018). Adaptive scheduling in smart factories: Balancing flexibility and throughput. *Flexible Services and Manufacturing Journal*, 32(1), 42–58.

Chapter 10

Applications of additive manufacturing across agricultural, aviation, automotive, medical, food processing, and sports industries

S. Syath Abuthakeer, K. Deepak Arjun, R. S. Raghul, P. B. Mithiran, and Jaffar Ali

10.1 AGRICULTURAL INDUSTRY

10.1.1 Introduction

Additive manufacturing (AM) plays a vital role in the globalized world as it provides innovative technologies to manufacture customized parts with different materials in different volumes [1–3]. Food and agriculture sector needs extensive customization toward designing and development of types of equipment. AM, also known as three-dimensional (3D) printing, is emerging as a transformative tool in agriculture. By enabling the rapid and cost-effective production of customized tools, equipment, and components (Figure 10.1), AM addresses specific challenges faced by farmers and agribusinesses. From precision farming tools to advanced irrigation systems, AM offers a new paradigm for the design and manufacturing of agricultural technologies [4].

10.1.2 Role of additive manufacturing

10.1.2.1 Precision farming tools

Precision farming aims to optimize agricultural inputs like water, fertilizer, and pesticides, reducing waste and improving yield. AM plays a crucial role by enabling the production of custom sensors, drone parts, and GPS-guided tools. For instance, 3D-printed soil moisture sensors can be designed to specific field requirements, allowing farmers to monitor and manage irrigation efficiently [5, 6]. Similarly, AM is used to create drone components for aerial mapping and crop health monitoring, ensuring that resources are used judiciously.

DOI: 10.1201/9781003569633-10

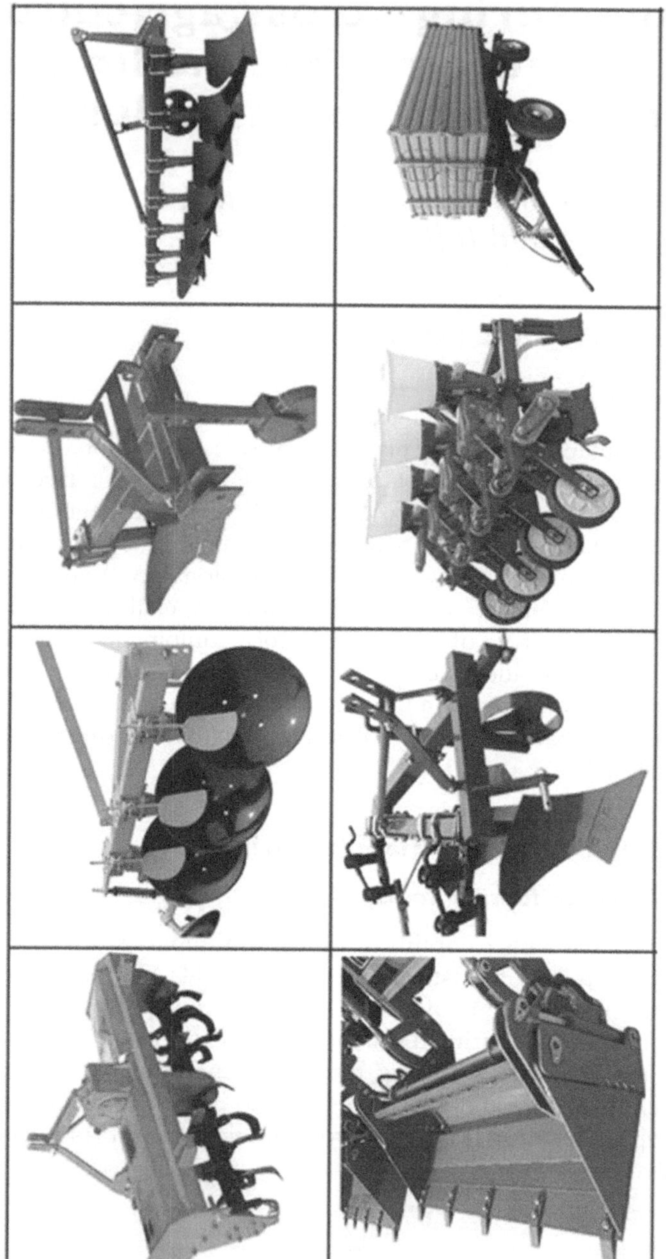

Figure 10.1 Agriculture tools.

10.1.2.2 Irrigation systems and components

Efficient irrigation is essential for sustainable agriculture, especially in regions facing water scarcity. AM allows for the creation of customized sprinkler nozzles, drip irrigation emitters, and filtration systems that optimize water distribution. For example, researchers have developed 3D-printed irrigation emitters with micro-scale channels that improve water flow and reduce clogging. These innovations ensure precise water delivery to crops, minimizing waste and enhancing yield.

10.1.2.3 Equipment prototyping and spare parts

Farm machinery often operates under harsh conditions, leading to wear and tear. AM enables the rapid prototyping of agricultural equipment, such as plows, seeders, and harvesters, allowing manufacturers to test and refine designs quickly. Additionally, AM provides farmers with a cost-effective way to produce spare parts on demand.

10.1.2.4 Customized greenhouse components

Greenhouses play a critical role in modern agriculture by enabling controlled-environment farming. AM allows for the production of custom greenhouse components (Figure 10.2), such as plant trays, irrigation systems, and structural connectors. For example, 3D-printed shading systems can be designed to regulate light and temperature, creating optimal conditions for crop growth [7]. This capability is particularly beneficial for high-value crops like fruits, vegetables, and flowers.

10.1.3 Case studies and application

10.1.3.1 3D-printed hydroponic systems

Hydroponics, a soil-less farming method, is gaining popularity for its ability to grow crops in limited spaces with minimal water usage. AM has been used to create modular hydroponic systems that are lightweight, scalable, and customizable [8]. For instance, a startup in the Netherlands developed 3D-printed vertical farming modules that maximize space utilization and improve water circulation.

10.1.3.2 John Deere: Custom spare parts on demand

John Deere, a global leader in agricultural machinery, has embraced AM to provide spare parts to its customers. Farmers can order 3D-printed

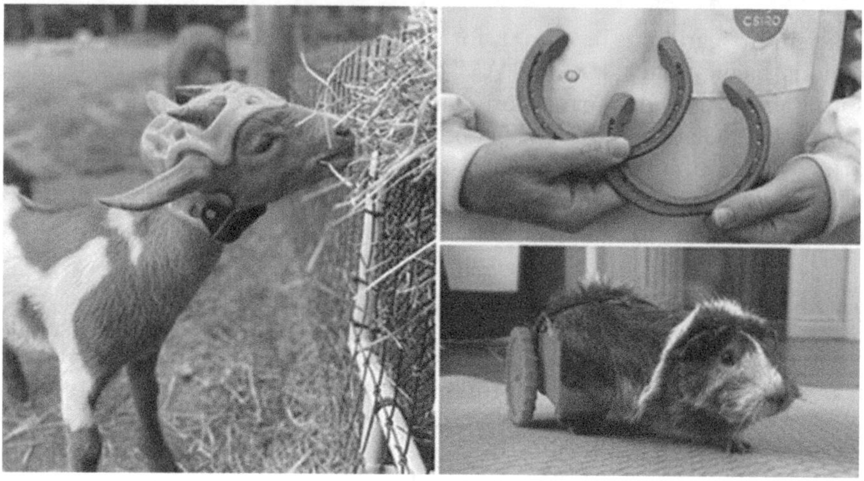

Figure 10.2 3D-printed animal prosthetics.

components, such as tractor gears and hydraulic fittings, through the company's online platform.

10.1.3.3 Prototyping of autonomous agricultural robots

Agricultural robotics is a rapidly growing field, with AM playing a crucial role in prototyping and manufacturing robot components. For example, a Japanese robotics company used AM to develop autonomous weed removal robots equipped with advanced sensors. The robots' lightweight and durable 3D-printed parts enabled faster field deployment and reduced costs.

10.1.3.4 3D-printed poultry farming tools

3D-Printed Beehives for Sustainable Beekeeping

In poultry farming, AM has been used to create custom egg trays, feeders, and water dispensers [9]. A poultry farm in Brazil implemented 3D-printed feeders with improved designs that minimized feed wastage and reduced maintenance efforts. These tools have enhanced operational efficiency and profitability.

10.1.4 Challenges

Despite its potential, AM in agriculture faces several challenges. One significant issue is the durability and longevity of 3D-printed components,

Figure 10.3 Aircraft manufacturing industry.

as agricultural equipment often operates in demanding environments. Materials used in AM must withstand extreme temperatures and mechanical stress, which can be challenging with current technologies.

10.1.5 Future trends

The future of AM in agriculture is promising, with advancements in materials, software, and manufacturing processes driving innovation. Biodegradable and composite materials are being developed to create eco-friendly 3D-printed agricultural products [10]. For example, researchers are exploring the use of plant-based polymers for printing tools and components, reducing the industry's carbon footprint.

10.2 AVIATION INDUSTRIES

10.2.1 Introduction

The aviation industry is a vital global sector, connecting people, goods, and economies across the world. It is characterized by stringent safety requirements, high costs, and the need for constant innovation. With the International Air Transport Association (IATA) projecting a 3.4% annual growth in global air traffic through 2037, the demand for more efficient, lightweight, and sustainable aircraft is increasing. However, traditional manufacturing processes often struggle to meet the demands for customization, reduced weight, and complex geometries, particularly in aerospace components (Figure 10.3).

10.2.2 Role of additive manufacturing

10.2.2.1 Lightweight components for fuel efficiency

Fuel efficiency is a critical factor in aviation, as fuel costs account for a significant portion of an airline's operating expenses. AM allows for the production of lightweight components that maintain strength and durability while reducing overall weight [11, 12].

10.2.2.2 Rapid prototyping and design iteration

The development of new aircraft involves extensive testing and prototyping to meet safety and performance standards. AM accelerates this process by enabling rapid prototyping of complex components. Engineers can quickly iterate designs, test prototypes, and incorporate feedback, significantly reducing development time.

10.2.3 Case studies and applications

10.2.3.1 GE Aviation: 3D-printed fuel nozzle

GE Aviation's development of a 3D-printed fuel nozzle (Figure 10.4) for its LEAP engine is one of the most celebrated examples of AM in aviation. By consolidating 20 parts into a single unit, AM reduced the nozzle's weight and increased its durability. The component's innovative design also improved fuel atomization, enhancing engine performance and reducing emissions.

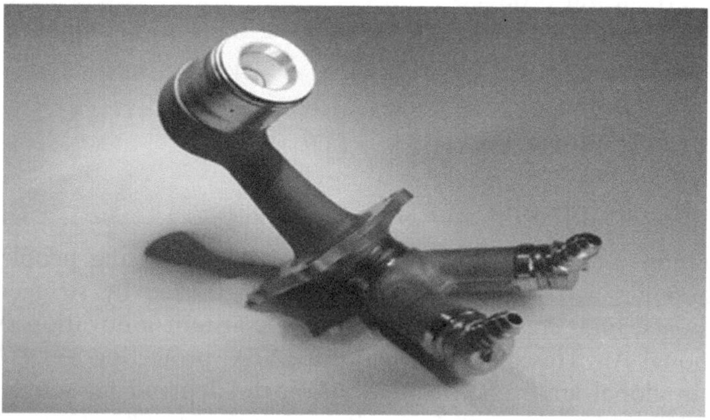

Figure 10.4 3D-printed nozzle.

10.2.3.2 Rolls-Royce: Additive manufacturing for engine blades

Rolls-Royce has utilized AM to manufacture turbine blades for its Trent XWB engines, used in the Airbus A350. These blades (Figure 10.5), produced using laser powder bed fusion (LPBF), feature advanced cooling channels that enhance thermal efficiency and reduce wear [13, 14]. The ability to produce such complex geometries would be impossible with traditional methods, highlighting AM's transformative potential in engine manufacturing.

10.2.3.3 NASA: Additive manufacturing in space exploration

While not strictly commercial aviation, NASA's use of AM for aerospace applications highlights its potential. NASA has developed 3D-printed rocket engine components, such as injectors and nozzles, which can withstand extreme pressures and temperatures. These advancements have implications for both space exploration and next-generation aircraft propulsion systems.

Figure 10.5 Rolls Royce fan blades.

10.2.4 Challenges

Despite its advantages, the adoption of AM in aviation is not without challenges. One of the primary issues is the stringent regulatory requirements for aerospace components.

10.2.5 Future trends

The future of AM in aviation is promising, with advancements in materials, software, and hybrid manufacturing techniques paving the way for broader adoption. Researchers are exploring the development of high-performance polymers and composites that combine lightweight properties with exceptional strength and durability.

10.3 AUTOMOTIVE INDUSTRIES

10.3.1 Introduction

The automobile industry is recognized as one of the most influential sectors shaping global economies, societies, and individual lifestyles. Therefore, fierce competition among different companies is continuously undergoing, and special attention is focused on innovations to improve competitiveness. In the past several years, AM has emerged as an innovative technology in applications in the automobile industry with significant advantages over traditional techniques [15]. Consequently, increasing efforts have been paid to combine AM technology with the development of the automobile industry (Figure 10.6). Currently, many automobile players are optimizing their industrial layout by incorporating innovative AM techniques, and meanwhile, a lot of research progress has been achieved to meet the market demand.

10.3.2 Role of additive manufacturing

AM plays a multifaceted role in the automotive industry, reshaping the way vehicles are designed, prototyped, and manufactured. Unlike traditional subtractive manufacturing, which involves cutting away material from a solid block, AM builds parts layer by layer, enabling unmatched design flexibility and material efficiency.

10.3.2.1 Production of lightweight components

One of the most significant contributions of AM to automotive manufacturing is the production of lightweight components. Using lattice structures and topology optimization, AM reduces the weight of parts like brackets, mounts, and suspension systems without compromising strength or durability.

Figure 10.6 Tesla manufacturing industry.

10.3.2.2 Customization and personalization

With AM, automakers can offer unprecedented levels of customization. Interior components, such as dashboards, door trims, and gear knobs, can be tailored to individual customer preferences (Figure 10.7) [16]. Luxury brands like Bugatti and Rolls-Royce use AM to deliver bespoke designs that enhance the exclusivity of their vehicles. BMW's MINI series, for example, offers customers the ability to personalize their car interiors with unique patterns and text, all made possible by AM.

10.3.2.3 Production of high-performance engine components

AM is being used to create highly efficient engine components such as pistons, cylinder heads, and turbocharger housings. For example, Porsche has utilized laser- based AM to manufacture pistons for its high-performance engines. These 3D-printed pistons feature optimized cooling channels, allowing for better thermal management and higher engine performance [17]. This innovation has enabled Porsche to improve engine efficiency by up to 30%, showcasing AM's ability to deliver performance-critical parts.

10.3.3 Case studies and applications

10.3.3.1 Ford Motor Company

Ford has been at the forefront of AM adoption, using the technology for prototyping, tooling, and end-use parts. One notable application is the development

Figure 10.7 Repairing classic Volvo parts using 3D printing.

of lightweight brake brackets for the Ford GT. By using AM, Ford reduced the weight of these components by 35%, improving overall vehicle performance. Ford also uses AM to produce customized tools for its assembly lines, cutting lead times by 50% and saving millions of dollars annually.

10.3.3.2 Volkswagen

Volkswagen's use of AM at its Wolfsburg plant exemplifies how 3D printing can enhance manufacturing efficiency. The company produces jigs, fixtures, and gauges using AM, which are essential for vehicle assembly.

10.3.3.3 Rolls-Royce: Bespoke interior features

Rolls-Royce has embraced AM to deliver bespoke interior features for its luxury vehicles. From personalized trim accents to intricate grille designs, AM enables Rolls- Royce to offer its clients unparalleled levels of customization. This not only enhances customer satisfaction but also reinforces the brand's exclusivity.

10.3.3.4 Daimler Trucks: AM for spare parts on demand

Daimler Trucks employs AM to produce spare parts for its commercial vehicles. By adopting a digital inventory approach, Daimler can produce parts such as pump covers, air nozzles, and brackets on demand. This eliminates the need for physical storage, reduces lead times, and lowers logistics costs.

10.3.4 Challenges

One of the primary challenges is material performance. Although AM materials have advanced significantly, many still lack the durability, heat resistance, and impact strength required for critical automotive applications. For instance, while polymers like acrylonitrile butadiene styrene (ABS) and polylactic acid (PLA) are suitable for prototyping, they often fall short for end-use parts that require long-term reliability.

10.3.5 Future trends

Looking forward, the integration of advanced software tools like generative design with AM is expected to unlock new possibilities. Generative design algorithms optimize part geometry for specific performance criteria, enabling the creation of ultra-lightweight and efficient components. When combined with AM, these designs can be realized with minimal material waste.

10.4 MEDICAL INDUSTRIES

10.4.1 Introduction

The medical industry (Figure 10.8) is one of the most critical sectors of the global economy, responsible for improving and preserving human life through innovative solutions. With an aging global population and rising incidences of chronic diseases, the demand for advanced medical devices, implants, and surgical tools has grown exponentially. According to the World Health Organization (WHO), the global healthcare market was valued at over $10 trillion in 2022, with medical devices and equipment accounting for a significant portion. However, this rapid growth also presents challenges, such as personalized patient care, affordability, and the need for precise and minimally invasive treatments.

10.4.2 Role of additive manufacturing

10.4.2.1 Patient-specific implants and prosthetics

One of the most impactful applications of AM in the medical field is the production of patient-specific implants and prosthetics. Unlike traditional manufacturing, AM can create highly customized solutions tailored to the unique anatomy of each patient. For example, titanium cranial implants

Figure 10.8 Aerial view of Bayer Pharmaceutical Industry.

manufactured using direct metal laser sintering (DMLS) offer a perfect fit and biocompatibility, reducing surgery time and improving patient outcomes.

10.4.2.2 Dental applications

AM has become a cornerstone of modern dentistry, revolutionizing the production of crowns (Figure 10.9), bridges, dentures, and orthodontic aligners. Digital workflows, combined with AM technologies like selective laser sintering (SLS) and digital light processing (DLP), enable the rapid creation of precise dental components. Companies like Align Technology, the maker of Invisalign, leverage AM to produce millions of custom aligners annually.

10.4.3 Case studies and applications

10.4.3.1 Johnson & Johnson: Custom orthopedic implants

Johnson & Johnson's DePuy Synthes division uses AM to manufacture customized orthopedic implants. For example, the ATTUNE Knee System

Figure 10.9 Tooth Crowns made by SLS method.

employs AM to create implants tailored to individual patient anatomies, resulting in better joint alignment and reduced pain. This approach has significantly improved patient satisfaction and long-term outcomes in knee replacement surgeries.

10.4.3.2 Stryker: 3D-printed hip implants

Stryker, a global leader in medical technology, has adopted AM to produce hip implants with porous structures that promote bone integration. Using electron beam melting (EBM), Stryker has developed a range of implants that mimic natural bone properties, enhancing stability and reducing the risk of implant failure.

10.4.4 Challenges

Despite its transformative potential, AM in the medical industry faces several challenges [18]. Regulatory hurdles remain a significant barrier, as 3D-printed medical devices must undergo rigorous testing and approval processes to ensure safety and efficacy.

10.4.5 Future trends

The future of AM in the medical industry is promising, with ongoing research driving innovation. Advances in material science are expected to yield new biocompatible and bioactive materials, expanding the range of applications for AM.

10.5 FOOD PROCESSING INDUSTRIES

10.5.1 Introduction

Food processing industries are integral to the global food supply chain, bridging the gap between agricultural production and consumers. They involve a series of operations that transform raw agricultural materials—such as grains, fruits, vegetables, dairy, meat, and seafood—into consumable, shelf-stable products.

10.5.1.1 Primary processing

This involves basic preparation, like cleaning, sorting, grading, and milling. For example, milling wheat into flour or cleaning and packaging fresh fruits and vegetables for retail.

10.5.1.2 Secondary processing

Here, raw ingredients are combined and processed into intermediate or finished products, such as baking bread, fermenting yogurt, or producing canned goods.

10.5.1.3 Tertiary processing

This stage focuses on creating ready-to-eat or convenience foods, such as frozen meals, snacks, or pre-packaged products.

10.5.2 Challenges

Despite their benefits, food processing industries (Figure 10.10) face challenges such as high energy consumption, waste management, and maintaining nutritional quality during processing.

This sector is also rapidly evolving, integrating technologies like Internet of Things (IoT), artificial intelligence (AI), and blockchain for traceability, automation, and efficient supply chain management. Such advancements ensure transparency and enhance trust among consumers.

- **Selective laser sintering** is also being explored for its potential in creating novel food textures by sintering edible powdered materials using lasers (Figure 10.11). This method can produce layered confections with unique mouthfeels, expanding the possibilities for culinary innovation.
- **Ultrasound-assisted processing** is another innovative technology, utilizing high-frequency sound waves for various applications. It aids in flavor and oil extraction, tenderizes meat by breaking down

Figure 10.10 3D food printer and the final product.

Figure 10.11 Food processing through binder jetting.

fibers, and serves as an effective method for cleaning and decontaminating produce.

10.5.3 Case studies

10.5.3.1 Transforming bakery production lines

A bakery faced challenges with a critical component in their production line that frequently failed, leading to costly downtime. Collaborating with K3D, a company specializing in metal 3D printing, they redesigned the part using food-safe AM materials.

10.5.3.2 Enhancing meat processing efficiency

The Australian Meat Processor Corporation (AMPC) sought to innovate the red meat processing industry by integrating 3D printing technologies. By adopting metal AM, they developed customized tools and components tailored to specific processing needs.

10.5.3.3 3D-printed food for dysphagia patients

Researchers at the University of the West of England, Bristol, developed 3D-printed food tailored for individuals with swallowing difficulties (dysphagia). Using pureed ingredients like garden peas and Greek yogurt, they created nutrient-rich, caloriedense meals that are easy to chew and swallow.

10.5.4 Challenges

The food industry is currently navigating a complex landscape of challenges that impact production, distribution, and consumer satisfaction. One significant issue is supply chain disruptions, which have been exacerbated by global events, leading to delays and increased costs.

10.5.5 Future trends

10.5.5.1 Sustainability and upcycled ingredients

Consumers are increasingly drawn to eco-friendly products that minimize waste and environmental impact. Upcycled ingredients—those derived from by-products of food production—are gaining traction as a sustainable alternative. These not only reduce food waste but also align with growing environmental consciousness.

10.5.5.2 Plant-based and alternative proteins

The shift toward plant-based diets continues to gain momentum, driven by health, ethical, and environmental concerns. Innovations in plant-based meat, dairy, and seafood alternatives are expanding rapidly. Additionally, lab-grown or cultured meats are emerging as a promising solution for meeting protein demands without the environmental impact of traditional livestock farming.

10.6 SPORTS INDUSTRIES

10.6.1 Introduction

The sports industry is a vibrant and rapidly growing sector that spans the design, production, and distribution of a wide array of products, including equipment, apparel, and footwear. This industry caters to professional athletes, recreational players, and fitness enthusiasts alike.

Manufacturing in the sports industry begins with the selection of raw materials tailored to the specific requirements of the product. Textiles and fabrics, for example, are essential in creating apparel and footwear that are durable, breathable, and moisture-wicking. Metals and composites, such as carbon fiber and aluminum, are commonly used in sports equipment like bicycles, tennis rackets, and golf clubs for their lightweight yet strong properties. Polymers and plastics find applications in protective gear, balls, and synthetic turf, offering versatility and resilience.

10.6.2 Challenges

The sports industry has been facing several challenges in recent times, driven by both internal and external factors. One significant issue is **supply chain disruptions,** exacerbated by global events like the COVID-19 pandemic and geopolitical tensions. These disruptions have led to delays in production, rising material costs, and difficulties in sourcing components, affecting the availability and cost of sports products.

10.6.3 Role of additive manufacturing

10.6.3.1 Customization and personalization

AM enables the creation of personalized sports equipment, such as custom-fit helmets, footwear, or prosthetics. Athletes can benefit from products tailored to their specific needs, enhancing performance and comfort. For example, custom shoes can be designed based on an athlete's foot structure and specific performance demands.

Figure 10.12 Fused deposition modelling.

10.6.3.2 Lightweight and high-performance components

AM allows the production of lightweight yet strong components for sports equipment. Materials like carbon fiber composites can be used to create high-performance parts for bicycles, tennis rackets, or golf clubs. These components offer strength without adding weight, improving the overall performance of the equipment.

10.6.4 Additive manufacturing technologies

10.6.4.1 Fused deposition modelling (FDM)

FDM (Figure 10.12) is one of the most common 3D printing methods used in the sports industry. It involves melting thermoplastic material, such as PLA or ABS, and extruding it layer by layer to build the final object.

10.6.4.2 Selective laser sintering

SLS is another popular 3D printing method used in the sports industry, particularly for creating complex parts from materials like nylon, metals, and composites. It is ideal for producing strong, durable parts for sports equipment such as lightweight frames for bicycles, golf clubs, and protective gear.

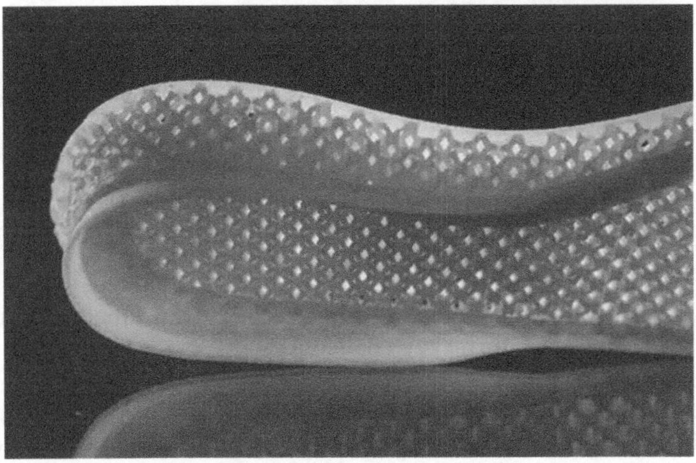

Figure 10.13 Shoe made using stereolithography method.

10.6.4.3 Stereolithography (SLA)

SLA uses ultraviolet light to cure liquid resin layer by layer to form high-precision parts. This technology is often used for prototyping high-end sports products that require fine details, such as customized helmets or racing car components, where precision and finish quality are critical.

10.6.5 Case studies

10.6.5.1 Custom running shoes by Adidas

Adidas has pioneered the use of 3D printing in the production of custom running shoes (Figure 10.13). The company introduced the **Adidas Future craft 4D**, a running shoe with a 3D-printed midsole made from a material called **Light strike**. Using **Digital Light Synthesis** (DLS) technology, a form of **SLA**, Adidas creates midsoles with precise geometries tailored to an athlete's specific needs.

10.6.5.2 3D-printed helmets by Riddell

Riddell, a leading manufacturer of American football helmets, has embraced **additive manufacturing** to create custom-fit helmets that enhance player safety. Using **3D scanning** and **3D printing** technologies, Riddell has been able to create personalized helmet liners that provide better protection against impacts.

10.6.5.3 Formula 1 Racing Parts by Mercedes-Benz

In Formula 1, teams like Mercedes-Benz are using **AM** to produce lightweight, high-strength parts for their race cars. Using **metal 3D printing** techniques such as **DMLS**, Mercedes creates components such as brake ducts, engine parts, and cooling systems that are not only lighter but also stronger than traditionally manufactured parts.

10.6.6 Challenges

The sports industry has been facing several challenges in recent years, driven by evolving market demands, technological advancements, and global events. Here are some of the key challenges:

- *Supply chain disruptions:* One of the most significant challenges faced by the sports industry in recent times is supply chain disruptions. The COVID-19 pandemic, combined with ongoing global trade issues and logistics challenges, has led to delays in production, higher transportation costs, and shortages of raw materials. These disruptions affect the timely delivery of sports equipment, apparel, and footwear, causing inventory shortages and negatively impacting sales.
- *Increasing demand for sustainability*: There is a growing demand from consumers for more sustainable products in the sports industry.

Athletes and enthusiasts are becoming more environmentally conscious, seeking sportswear, equipment, and footwear made from recycled materials or produced through sustainable manufacturing practices. Meeting these demands while maintaining high product performance and quality is a complex task.

10.6.7 Future trends

The sports industry is continuously evolving, driven by technological advancements, changing consumer behaviors, and new global challenges. Several future trends are shaping the direction of the industry. Here are some key trends to watch for in the coming years.

10.6.7.1 Integration of technology and wearables

The use of wearable technology in sports is expected to continue growing. Devices like fitness trackers, smartwatches, and performance monitoring systems are becoming more sophisticated, offering real-time insights into an athlete's performance, health metrics, and recovery. Future advancements may lead to even more seamless integration of sensors into clothing, shoes, and equipment, allowing for continuous data collection without the need for bulky devices.

Moreover, smart fabrics embedded with sensors will enable real-time tracking of metrics such as heart rate, temperature, and movement patterns, offering deeper insights for both professional athletes and casual fitness enthusiasts.

10.6.7.2 Additive manufacturing (3D printing)

AM is poised to play an even larger role in the sports industry, particularly in the areas of product customization and prototyping. The ability to create lightweight, high-strength components tailored to an athlete's specific needs is a game changer. Expect to see even more advanced uses of 3D printing in creating customized footwear, protective gear, prosthetics, and sports equipment. As the technology becomes more accessible, it may also lead to on-demand manufacturing, reducing waste and supply chain issues.

10.6.7.3 Sustainability and eco-friendly products

Sustainability will remain a key focus in the future of the sports industry. Consumers are increasingly demanding eco-friendly products, and brands are responding by adopting sustainable materials and manufacturing processes. This includes the use of recycled materials in sports apparel and equipment, as well as the adoption of biodegradable products. Companies may also explore the circular economy, where products are designed to be reused or recycled, minimizing their environmental impact.

10.6.7.4 Virtual and augmented reality

The use of virtual reality (VR) and augmented reality (AR) is transforming both training and fan engagement. VR is already being used to simulate training environments, helping athletes practice in controlled yet dynamic settings. In the future, VR could be used for remote training, allowing athletes to experience competitive scenarios without needing to travel.

AR, on the other hand, is enhancing the fan experience, with interactive displays at live events, real-time stats overlays, and immersive viewing experiences. AR-enabled apps may also allow fans to interact with sports content in new and exciting ways, such as virtually trying on merchandise or viewing replays in 3D.

10.6.7.5 E-sports and digital engagement

E-sports is rapidly becoming a major segment of the sports industry, with competitive gaming attracting millions of fans globally. The integration of e-sports with traditional sports events, such as virtual leagues or gaming tournaments, is expected to grow. As younger generations engage with e-sports, traditional sports organizations are exploring ways to merge physical and digital sports experiences.

10.6.7.6 Data analytics and performance optimization

The use of big data and AI in sports will continue to grow. Teams and athletes are increasingly relying on data analytics to optimize performance, track injuries, and design customized training regimens. Predictive analytics will help prevent injuries by identifying patterns that could lead to strain or fatigue. AI-driven coaching tools will provide real-time feedback, allowing athletes to adjust their performance on the fly.

10.6.7.7 Globalization of sports

As the global fanbase for sports continues to grow, the sports industry will increasingly look to tap into emerging markets in regions such as Asia, Africa, and Latin America. The rise of digital streaming platforms and online sports betting will further globalize the reach of sports, allowing fans to follow their favorite teams and athletes from anywhere in the world.

10.6.7.8 The rise of hybrid and remote sports events

The way sports events are conducted and consumed is changing. The hybrid event model, which blends in-person and virtual experiences, will likely become a permanent trend. Fans who cannot attend in person will have access to live-streamed content, VR/AR experiences, and interactive platforms, creating a more inclusive global audience.

10.7 CONCLUSION

AM has emerged as a revolutionary technology across multiple industries, including automotive, medical, agriculture, and aviation. Each sector demonstrates the versatility and transformative potential of AM, from enhancing efficiency and customization to reducing costs and promoting sustainability.

In the **automotive industry**, AM has become a critical tool for lightweighting, rapid prototyping, and mass customization. By enabling the production of complex geometries and integrating components into single structures, AM has significantly enhanced the performance and fuel efficiency of vehicles

In the **medical field,** AM has redefined the possibilities for personalized healthcare. Custom implants, prosthetics, and surgical tools demonstrate its unparalleled ability to meet individual patient needs. The rise of bioprinting has further expanded the potential of AM, paving the way for breakthroughs in organ regeneration and tissue engineering.

In **agriculture,** AM addresses the pressing need for resource efficiency and innovation in farming practices. From custom irrigation systems and precision tools to hydroponic systems and livestock management solutions, AM provides tailored, cost- effective solutions. Its ability to produce on-demand spare parts and tools offers a lifeline to farmers in remote areas, promoting self-reliance and minimizing downtime.

In the **aviation industry,** AM has transformed the production of lightweight components, enabling more fuel-efficient and sustainable aircraft. By facilitating the creation of complex, high-performance parts, AM addresses the stringent safety and performance requirements of aerospace engineering.

In conclusion, the application of AM across the automotive, medical, agriculture, and aviation industries highlights its transformative impact and vast potential. As the technology continues to evolve, it will not only redefine manufacturing but also contribute to a more sustainable and innovative future.

REFERENCES

1. Javaid, M., & Haleem, A. (2019). Using additive manufacturing applications for design and development of food and agricultural equipment. *International Journal of Materials and Product Technology*, 59(2–3), 225–238. https://doi.org/10.1504/IJMPT.2019.097662
2. Dizon, J. R. C., Espera, A. H., Chen, Q., & Advincula, R. C. (2018). 3D printing of polymers for food and agricultural applications: A review. *Journal of the Science of Food and Agriculture*, 98(15), 5555–5567. https://doi.org/10.1002/jsfa.9124
3. Karuthapandi, S., Ramu, M., & Thyla, P. R. (2017) Effects of the use of a flat wire electrode in gas metal arc welding and fuzzy logic model for the prediction of weldment shape profile, *Journal of Mechanical Science and Technology*, 31(5), 2477–2486.
4. Baig, M. M., & Muzammil, M. (2020). Additive manufacturing applications in agriculture: An overview. *Materials Today: Proceedings*, 26(2), 1571–1576. https://doi.org/10.1016/j.matpr.2020.02.315
5. Gibson, I., Rosen, D. W., & Stucker, B. (2015). Additive manufacturing technologies: 3D printing, rapid prototyping, and direct digital manufacturing. *Springer*. https://doi.org/10.1007/978-1-4939-2113-3
6. Frazier, W. E. (2014). Metal additive manufacturing: A review. *Journal of Materials Engineering and Performance*, 23, 1917–1928. https://doi.org/10.1007/s11665-014-0958-z

7. Herderick, E. D. (2011). Additive manufacturing of metals: A review. *Materials Science and Technology Conference and Exhibition*, 1, 1413–1425. https://pmc.ncbi.nlm.nih.gov/articles/PMC10744938/
8. Gao, W., Zhang, Y., & Ramanujan, D. (2015). The status, challenges, and future of additive manufacturing in engineering. *Computer-Aided Design*, 69, 65–89. https://doi.org/10.1016/j.cad.2015.04.001
9. Ford, S., & Despeisse, M. (2016). Additive manufacturing and sustainability: An exploratory study of the advantages and challenges. *Journal of Cleaner Production*, 137, 1573–1587. https://doi.org/10.1016/j.jclepro.2016.04.150
10. Berman, B. (2012). 3-D printing: The new industrial revolution. *Business Horizons*, 55(2), 155–162. https://doi.org/10.1016/j.bushor.2011.11.003
11. Ventola, C. L. (2014). Medical applications for 3D printing: Current and projected uses. *Pharmacy and Therapeutics*, 39(10), 704–711. www.ncbi.nlm.nih.gov/pmc/articles/PMC4189697/
12. Murphy, S. V., & Atala, A. (2014). 3D bioprinting of tissues and organs. *Nature Biotechnology*, 32, 773–785. https://doi.org/10.1038/nbt.2958
13. Lipton, J. I., Cutler, M., Nigl, F., Cohen, D., & Lipson, H. (2015). Additive manufacturing for the food industry. *Trends in Food Science & Technology*, 43(1), 114–123. https://doi.org/10.1016/j.tifs.2015.02.004
14. Xhameni, A., Cheng, R., & Farrow, T. (2022). A precision method for integrating shock sensors in the lining of sports helmets by additive manufacturing. *arXiv preprint arXiv:2206.11202*. Available at: https://arxiv.org/abs/2206.11202
15. Sybridge Technologies. (2022). How additive manufacturing transforms the sporting goods industry [Online]. Available at: https://sybridge.com/how-additive-manufacturing-transforms-sporting-goods-industry/
16. 3D Textiles. (2024). *Wikipedia*. Available at: https://en.wikipedia.org/wiki/3D_textiles
17. Ingle, S. (2025). A vision of sport in 2050: Robot leagues, chips in brains and players in their 50s. *The Guardian [online]*. Available at: www.theguardian.com/sport/2025/jan/06/a-vision-of-sport-in-2050-robot-leagues-chips-in-brains-and-players-in-their-50s
18. Karuthapandi, S., Murugan, R., & Palani, P.K. (2016) Study and analysis of the macrostructure characteristics in FCAW with the use of a flat wire electrode and by optimizing the process parameter using the Taguchi method and regression analysis, *High Temperature Material Processes*, 20(3), 6–10.

Chapter 11

Advancements in quality control and monitoring strategies for wire arc additive manufacturing

A comprehensive review

N. Ganesh and S. Ajithkumar

11.1 INTRODUCTION

An essential progression in traditional fabrication techniques is wire arc additive manufacturing (WAAM), which facilitates the creation of complex geometries with enhanced material properties and heightened operational efficiency. The implementation of comprehensive quality assurance and monitoring systems is imperative as WAAM technology expands across various industrial sectors to guarantee uniform output and dependability throughout the manufacturing continuum. Such technological advancements facilitate the swift identification of defects and the modification of manufacturing parameters as required, thereby improving product quality and minimizing material wastage. This review seeks to investigate the most recent developments in quality control methodologies and monitoring systems specifically designed for WAAM, underscoring their importance in bolstering process stability and product integrity. WAAM is an advanced technique in metal three-dimensional (3D) printing that employs an electric arc to melt metal wire and apply it to a substrate, thus allowing for the meticulous formation of complex shapes over time. This technology is valued for its economic feasibility, high material efficiency, and versatility in material composition, making it suitable to produce large-scale and complex industrial components, notably in the aerospace, automotive, and marine engineering sectors [1–4]. WAAM uses traditional welding procedures like gas metal arc welding (GMAW) and is frequently automated using computer-aided design/computer-aided manufacturing (CAD/CAM) systems on robotic or computer numerical control (CNC) platforms, which allows for fine control over the deposition process [5, 6]. The process parameters, such as wire feed rate, welding speed, and gas flow rate, play an important role in defining final product quality, impacting characteristics such as dimensional accuracy, mechanical qualities, and defect development [7]. For example, improving these parameters can dramatically minimize gas entrapment while increasing tensile strength and elastic modulus in aluminum alloys [2]. The microstructure of WAAM-fabricated components can vary substantially,

DOI: 10.1201/9781003569633-11

with differences in solidification rates affecting mechanical characteristics such as tensile and yield strength [8]. Furthermore, anisotropy in mechanical characteristics, which is frequently found due to columnar grain formation along the construction direction, may be controlled by process optimization and heat treatment [4, 9]. The multifaceted nature of WAAM is significantly augmented by the utilization of open-source software frameworks that facilitate the integration with diverse robotic and sensor systems, thereby permitting real-time modifications to processes and the implementation of quality control measures [6]. Empirical comparative analyses have indicated that the utilization of metal-cored wires leads to a reduction in manufacturing duration and an enhancement in deposition rates in contrast to solid wires, thus illustrating the economic advantages of WAAM within industrial contexts [10]. Ultimately, WAAM represents a groundbreaking method in additive manufacturing, offering enhanced productivity, economic viability, and the ability to create intricate metal components of outstanding quality.

Quality assurance and monitoring in WAAM are critical for ensuring the production of outstanding components, particularly given the process's inherent instability and complexity. The importance of these protocols is heightened by the need to maintain geometric accuracy and material properties, which are frequently threatened by abnormalities such as spatter formation and oxygen incorporation throughout the manufacturing process [11, 12]. In the aerospace industry, for example, the production of large components using alloys such as Ti-6Al-4V necessitates stringent quality control measures to avoid complications such as embrittlement caused by oxygen absorption, which can have a significant impact on the mechanical properties of the final product [12]. The objectives of quality assurance in WAAM encompass the stabilization of the welding procedure through the implementation of real-time monitoring and control techniques, including the utilization of neural networks for semantic segmentation alongside conventional image processing methodologies. These techniques facilitate closed-loop control, thereby enhancing the geometric precision and mechanical properties of the produced components [11]. The scope of quality control includes the development of automated systems such as iWAAM (intelligent WAAM), which integrate process monitoring with control frameworks to regulate variables such as arc current and wire feed speed, improving the precision and repeatability of the production process [13]. The incorporation of advanced monitoring systems, including optical and infrared methods for non-destructive testing (NDT), enables high-resolution inspections and real-time feedback, which are crucial for maintaining the integrity of complex structures in industries such as aerospace and aviation [14]. The utilization of in situ sensors and cameras for the collection and analysis of real-time data in additive manufacturing is increasing, as it supports the identification and resolution of problems during the printing procedure [15]. To tackle the challenges of achieving flawless manufacturing and to guarantee the required consistency and dependability in WAAM processes, it is crucial to implement comprehensive quality control and monitoring systems [16–18].

11.2 FUNDAMENTALS OF WAAM

One of the most well-known methods in the field of additive manufacturing is WAAM, which uses electric current to provide heat for melting wire and layer-by-layer deposition to produce 3D metal parts. Similar to welding, this technique has several advantages for creating complex, large-scale structures with high material economy and quick prototyping [19]. In terms of materials, WAAM is very versatile, often using steel, nickel-based superalloys, and aluminum alloys, each selected for their own mechanical characteristics and particular application requirements [20, 21]. Key process parameters in WAAM include arc current, wire feed speed, travel speed, and shielding gas, all of which significantly influence the heat input and, consequently, the quality of the deposited parts [22]. To minimize flaws like residual stresses and deformation and to achieve desirable mechanical qualities like tensile strength and hardness, it is imperative that these parameters be optimized [23]. The aerospace, automotive, and marine engineering industries stand to gain from the cost-effectiveness, rapid deposition rates, and ability to produce large components offered by WAAM in comparison to other additive manufacturing techniques [24]. Nevertheless, challenges such as process stability and repeatability remain, highlighting the need for additional research and development to enhance its competitiveness and tackle issues like mechanical inconsistencies and microstructural defects [19, 23]. WAAM's place in contemporary production is being further cemented by the development of sophisticated algorithms and optimization techniques that increase accuracy, material efficiency, and heat control [25, 26]. Overall, WAAM stands out for its ability to produce near-net-shape components efficiently, though continuous advancements are essential to fully unlock its potential in the additive manufacturing landscape [23, 27]. The combination of machine learning (ML) and artificial intelligence (AI) within the design and manufacturing processes is anticipated to be crucial in addressing current challenges, facilitating the emergence of more creative applications and wider acceptance across multiple industries. This technological advancement not only improves the quality and dependability of produced components but also encourages an eco-friendlier strategy by reducing waste and energy usage throughout the manufacturing process.

11.3 CHALLENGES IN WAAM QUALITY CONTROL

WAAM presents several quality assurance problems, notably with product homogeneity and accuracy. The usefulness of created components can be considerably affected by fault detection, surface texture, dimensional precision, and material appropriateness. Figure 11.1 shows that the WAAM process's complexity—fast deposition rates, variable heat inputs, and thermal distortion—causes these difficulties. New monitoring methods, adaptive control mechanisms, and continual optimization are needed to increase product quality.

11.3.1 Common defects in WAAM (e.g., porosity, distortion, cracking)

WAAM is a promising method for generating large metallic components, but it is prone to a number of common faults that can significantly impact the mechanical properties and quality of the end product. Porosity is a prevalent defect in WAAM, which is frequently induced by process disruptions and irregularities during fabrication. It can substantially degrade mechanical performance, as seen in super duplex stainless steel, where porosity resulted in a considerable drop in yield strength, ultimate tensile strength, and fatigue strength, with fatigue fractures frequently arising from pores

Figure 11.1 Challenges in WAAM quality control.

[28]. Porosity also affects aluminum alloys, and it may be reduced using procedures such as interlayer friction stir processing, which improves mechanical characteristics by lowering porosity and refining grain size [29].

Another common defect in WAAM is cracking, which can result from the high residual stresses intrinsic to the process. These stresses result from the complicated heat cycles required in arc-based metallurgy, causing morphological errors and microstructural problems [30]. Another issue is distortion, which is frequently associated with heat input and deposition parameters and can result in deposition layer deviations and collapse [31]. Identifying and assessing these anomalies is critical, and sophisticated methodologies such as machine vision and thermal imaging have been developed to detect and quantify defects such as porosity and surface irregularities in real time, allowing for effective interlayer repairs and quality assurance [31, 32]. Moreover, alterations in production parameters, including wire feed velocity and deposition angle, may mitigate these imperfections by enhancing microstructural integrity and mechanical properties [33]. In summary, although WAAM offers substantial economic and efficiency advantages, it is imperative to address recurrent defects through enhanced detection methodologies and process optimization to produce high-quality components that are suitable for industrial applications [34].

11.3.2 Factors affecting build quality (e.g., heat input, deposition rate)

Build quality in WAAM is greatly influenced by a number of variables, including heat input, deposition rate, and process optimization techniques. Heat input is a crucial factor as it influences the mechanical characteristics, microstructure, and forming quality of the produced parts. As evidenced by the investigation into the characteristics of ZL205A components, it was determined that the minimal heat input corresponds with superior forming quality and enhanced mechanical properties; specifically, reduced heat input can lead to diminished porosity and augmented tensile strength [35]. Another significant factor to evaluate is the deposition rate; metal-cored wires exhibit superior effectiveness for industrial applications owing to their elevated deposition rates and diminished production times in contrast to solid wires [10]. In the pursuit of superior WAAM components, emphasizing the improvement of diverse process factors, including wire feed speed and temperature oversight, is critical. To augment deposition rates, material efficiency, and precision while concurrently minimizing expenditures, advanced algorithms such as Cost-Effective Heat Management and Adaptive Parameter Optimization are employed to dynamically adjust these variables [26]. The deposition methodology is further refined through the implementation of algorithms designed for real-time monitoring and control, including the Thermal Feedback Control Algorithm and the Adaptive

Deposition Control Algorithm, which collectively enhance surface quality and geometric accuracy [36]. In conclusion, the improvement of build quality in WAAM necessitates the integration of sophisticated process control methodologies, optimal heat input, and deposition rates [22].

11.3.3 Limitations of conventional inspection methods

Conventional inspection methods in WAAM face limitations due to challenges like high temperatures and surface irregularities inherent in the process. The high temperatures and surface irregularities inherent in WAAM complicate the inspection process, making it difficult to detect defects reliably during manufacturing [37]. Traditional ultrasonic methods work well for volumetric imaging but have trouble with the curved and non-planar shapes that are common in WAAM components, which can make it harder to find defects accurately [38]. Furthermore, the huge number of depositions needed in WAAM raises the possibility of defects forming; hence more advanced inspection methods are needed to guarantee quality. Although NDT techniques such as eddy current testing and phased-array ultrasonic techniques have been improved for WAAM, it is still difficult for them to detect all faults, particularly in large buildings. Furthermore, the implementation of real-time defect detection and in-line monitoring is critical, as these practices facilitate immediate modifications and reduce both waste and rework; however, conventional methodologies often inadequately address these requirements. Even with the significant advancements in NDT and in-line monitoring technologies such as laser Doppler vibrometry and X-ray fluorescence spectrometry, it is clear that more advancements are necessary to effectively address particular problems with WAAM. This suggests that standard inspection techniques alone are unable to ensure better quality of components produced with WAAM [39].

11.4 ADVANCES IN QUALITY CONTROL TECHNIQUES FOR WAAM

11.4.1 In situ monitoring

The methodologies for quality control in WAAM with in situ monitoring have notably advanced due to the incorporation of diverse innovative strategies and technological developments. Figure 11.2 shows the advances in quality control techniques for WAAM. One significant advancement is the use of adaptive control methods, which employ a wide range of algorithms to increase the control of layer thickness, improve the accuracy of thermal imaging, and strengthen the dependability of in situ monitoring, thus increasing the geometric accuracy and surface quality in WAAM operations [40]. Vision-based process monitoring has also been proposed, which uses

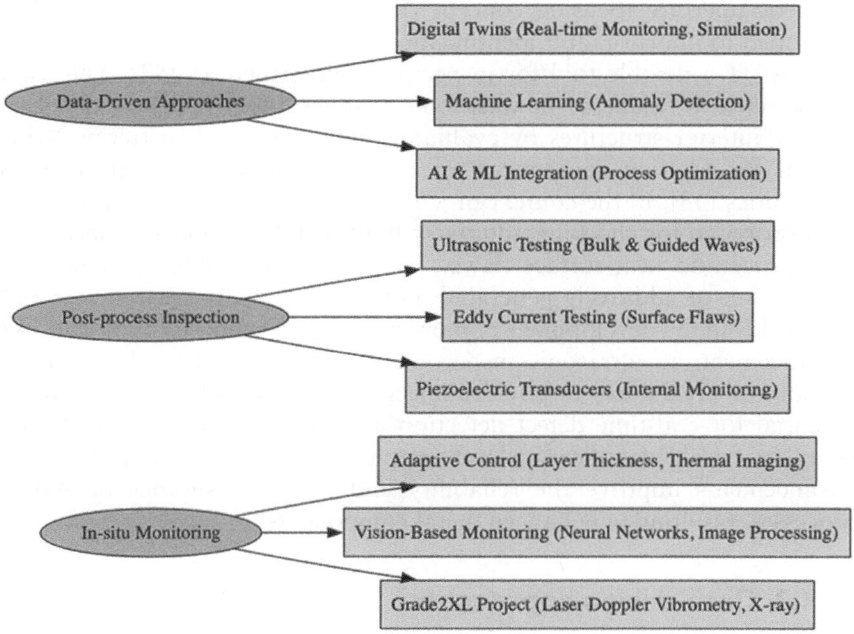

Figure 11.2 Advances in quality control techniques for WAAM.

neural networks for semantic segmentation to monitor the location of the welding wire and traditional image processing techniques to record spatter production, therefore improving process stability and geometric fidelity [11]. Furthermore, the Grade2XL project investigated NDT and in-line monitoring techniques including laser Doppler vibrometry and X-ray fluorescence spectrometry, which give real-time structural information and detect faults during the printing of extra-large structures [39]. Furthermore, to optimize WAAM, a complete framework including adaptive layer thickness management, in situ stress monitoring, and ML for process control was created, resulting in improvements in residual stress reduction, geometric accuracy, and process stability [41]. These advancements propel the evolution of WAAM technology, advancing the boundaries of additive manufacturing by enhancing quality and ensuring more reliable production processes.

11.4.2 Post-process inspection

WAAM has made tremendous progress in quality control approaches, notably with the use of NDT technologies such as ultrasonic testing (UT) and eddy current testing. UT is a well-established technology for identifying and characterizing material discontinuities that use both bulk and guided waves to provide high-resolution imaging of flaws. Techniques such as the synthetic aperture focusing concept (SAFT) and Lamb waves have been developed to improve fault identification and structural health monitoring, making UT a flexible tool for inspecting solid structures [42]. The use of piezoelectric transducers in ultrasonic systems improves the capacity to observe interior structures by evaluating reflections and scattering, which is vital for assuring component integrity in sectors such as nuclear and aeronautics [43]. In the context of WAAM, the eddy current approach has been proposed for checking additively built metal components, specifically for surface and near-surface flaws. This approach enables a quantitative comparison of additively generated and conventionally created materials, which aids in the development of more precise inspection technologies [44]. Advancements in ultrasonic inspection methodologies have been utilized to monitor procedures in powder injection molding (PIM), indicating the potential for real-time defect detection and process optimization through neural network classification and simplex control algorithms [45]. The advancements improve the reliability and quality assurance of WAAM processes, resulting in enhanced metal components.

11.4.3 Data-driven approaches

The strategies for ensuring quality in WAAM are progressively advancing with the integration of data-driven techniques. This evolution is especially apparent in the utilization of ML, AI, and digital twins (DTs). Contemporary studies have highlighted the critical role of DTs in the simulation and enhancement of manufacturing methodologies. They facilitate real-time oversight and regulation, thereby augmenting the precision and dependability of WAAM systems [46]. ML is employed within DTs to detect anomalies in sequential data inputs. This process safeguards the integrity of simulation models and enhances the efficacy of decision-making processes [47]. A DT-driven method for online quality control in the process industry has also been suggested. This method uses advanced algorithms, such as bidirectional gated recurrent units with attention mechanisms, to predict and improve product quality in real time [48]. A two-stage simulation-based method that uses DTs to produce data-driven control strategies complements this strategy and increases the adaptability and effectiveness of industrial processes [49]. In the context of food quality control, AI and ML are transforming traditional methods by enabling predictive analysis, which shifts the focus from reactive to proactive quality management, thereby reducing waste and ensuring compliance with safety standards [50]. Collectively, these innovations underscore the transformative potential of integrating AI, ML, and DTs in quality control, offering robust solutions for maintaining high standards in manufacturing and other industries.

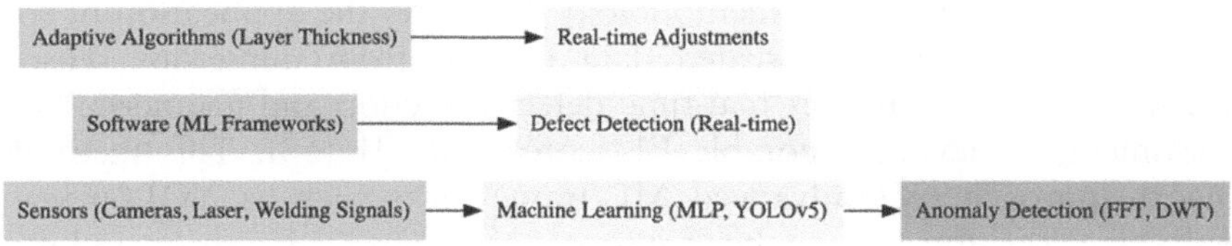

Figure 11.3 WAAM process monitoring strategies.

11.5 STRATEGIES FOR PROCESS MONITORING IN WAAM

11.5.1 Integration of sensors and IoT

Enhancing quality control and process optimization in the field of WAAM requires the integration of sensors and Internet of Things (IoT) for process monitoring. Figure 11.3 illustrates WAAM process monitoring strategies. To address the complexity inherent in WAAM, a variety of techniques have been examined, including the use of unsupervised ML algorithms for anomaly identification. These approaches, including isolation forest, local outlier factor, and one-class support vector machine, extract features in both the temporal and frequency domains. For example, fast Fourier transform (FFT) and discrete wavelet transform (DWT), which work in the frequency domain, make detection much more accurate [51, 52]. Adding a digital shadow approach that combines sensors from cameras, laser profilometers, and welding electric signal sensors makes monitoring systems even more flexible and 3D. Using ML methods such as multi-layer perceptron (MLP) and YOLOv5, this methodology identifies vulnerabilities and classifies them into different severity levels. The F1 score of 0.791 indicates that it is feasible to monitor in real time [53]. A further investigation into the connection between electrical signals and process instability is carried out with the assistance of multi-sensor systems. This shows how important it is to choose the right operating parameters to keep the process stable and reduce defects [54]. The detailed strategies described, such as ATM, MCA 2, ALTC 2, and RSM, stress how important it is for WAAM to change and get better over time. This makes it a feasible choice for producing massive metal components [55]. These studies indicate that combining sophisticated sensor technologies with the IoT in WAAM can improve process monitoring and control, potentially leading to higher deposition rates, surface quality, and material compatibility.

11.5.2 Role of software in monitoring strategies

In the field of WAAM, the significance of software is critical in advancing strategies for process supervision. This statement is especially pertinent when considering the implementation of advanced algorithms and

ML methodologies. For instance, the elucidation provided by Mu et al. regarding a digital shadow methodology illustrates the application of software to harmonize multi-scaled data derived from numerous sensors, which ultimately facilitates real-time defect detection and geometric error assessment, achieving a notable F1 score of 0.791 [53]. This methodological approach uses advanced ML frameworks such as YOLOv5 and multi-layer perceptrons to achieve its goals of effective supervision and well-informed decision-making. So, it highlights how crucial software is to the processing and understanding of complex data sets. Moreover, the study conducted by Abdullah et al. [39] regarding the incorporation of ML in process control highlights the essential function of software in the real-time modification of parameters such as layer thickness and stress distribution, improving the WAAM technique to attain greater precision and optimization. This increasing dependence on software to improve monitoring skills in WAAM is further highlighted by the utilization of deep learning algorithms for anomaly detection and surface roughness predictions, which have been investigated in a number of academic papers [56]. Al Hamed [57] elucidates that notwithstanding these advancements, challenges persist, particularly in the establishment of real-time micro-defect detection systems, suggesting that further research and software innovation are imperative for the enhancement of WAAM process monitoring in forthcoming endeavors. In the end, software plays a significant role in the development of WAAM as it enables complex monitoring and control systems, which in turn accelerate the technology toward further evolution.

11.5.3 Process feedback and adaptive control systems

In WAAM, efficient process oversight and adaptive control systems are vital for enhancing production and guaranteeing superior quality outputs. Numerous tactics have been formulated to satisfy these needs. One method involves employing adaptive algorithms, such as Adaptive Layer Thickness Control and Adaptive Deposition Control, which make real-time adjustments to layer thickness and deposition paths based on immediate feedback, thus improving geometric precision and reducing stress [36, 41]. Moreover, engaging ML algorithms, particularly multi-layer perceptrons and YOLOv5, enhances the pace of defect detection and decision-making processes, thereby increasing the flexibility of monitoring systems. Sensor fusion, which amalgamates information from welding electric signal sensors, cameras, and laser profilometers, elevates surveillance capabilities by synchronizing multi-dimensional data for thorough analysis [53]. Furthermore, the use of thermal feedback control algorithms and real-time monitoring systems ensures precise temperature management and process stability, both of which are critical for maintaining the integrity of produced components [36, 54]. Together, these tactics form a strong foundation for WAAM, allowing for dynamic modifications, continual improvements, and the creation of faultless components with improved surface quality and material compatibility [55, 54].

11.6 FUTURE DIRECTIONS AND RESEARCH OPPORTUNITIES

It is expected that future research activities for WAAM will focus on enhancing monitoring systems, particularly through the application of predictive analytics, in addition to addressing the economic and environmental ramifications of this emerging industry. Application of multi-sensor systems and frequency domain analysis are two sophisticated WAAM monitoring approaches that are essential for recognizing flaws such as porosity and arc instability, in addition to fluctuations in the process. Both these techniques are essential for controlling the process. The process employs both these methodologies. The DWT and FFT are two examples of the techniques that are employed in the analysis of welding signal data by these technologies. This facilitates the identification of anomalies in real time and enhances the overall reliability of the process. Predictive analytics integration, particularly using ML algorithms, is another fascinating area. By dynamically adjusting the layer thickness and stress distribution, these algorithms have the potential to enhance process parameters, improving both material integrity and geometric precision of the structure. Furthermore, WAAM exerts a considerable influence on both economic and ecological dimensions. In contrast to traditional manufacturing methods, including laser powder bed fusion (LPBF) and CNC milling, WAAM is renowned for its reduced environmental footprint and enhanced material efficiency. Because of its rapid deposition rates and low cost of feedstock materials, it provides economic benefits, particularly when it comes to the production of large components in batches. However, post-processing is required due to issues including poor surface smoothness and dimensional correctness, which can raise expenses and have an adverse effect on the environment [58]. To improve WAAM, future research should focus on additive component databases, hybrid manufacturing analysis, and AI [59, 60]. These projects will increase WAAM process efficiency and quality and broaden their application in automotive, marine, and aerospace engineering.

11.7 CONCLUSION

In conclusion, WAAM has substantial prospects for innovation across several sectors as it evolves, improving sustainability and operational efficiency while tackling existing challenges via focused research and development. Ultimately, these initiatives will herald a new epoch of sophisticated production systems that emphasize economic sustainability and environmental stewardship through the promotion of a more cohesive manufacturing framework. Recent advances in manufacturing technology will allow firms to adopt automation and real-time data analytics, creating production environments that are more responsive to market needs. As businesses embrace these technologies, WAAM will benefit from increased industry collaboration, creating a dynamic environment that fosters innovation and development. Beyond enhancing productivity, this collaborative environment will facilitate the exchange of best practices, which will ultimately yield a more resilient and sustainable manufacturing landscape capable of addressing future challenges. Consequently, industries will be better positioned to embrace the principles of circular economy, minimizing waste and maximizing resource efficiency while preserving competitive advantages in a marketplace increasingly oriented toward environmental considerations. The integration of advanced technologies will facilitate more intelligent manufacturing methodologies, empowering businesses to harness automation and data analytics for ongoing innovation and progress. In a swiftly changing industrial landscape, this transformation will foster long-term sustainability by nurturing a culture of adaptability that enables organizations to promptly adjust to market fluctuations and technological advancements.

REFERENCES

1. T. Abioye, B. D. Bankong, O. J. Ogundipe, O. O. Kusoro, and T. I. Ogedengbe, "A parametric study on the stability, geometry and hardness of AISI 308LSi WAAM-GMAW," *Materials Science Forum*, Nov. 2024, 1131, 49-60, doi: 10.4028/p-6mzfkt
2. M. S. Kumar, N. Jeyaprakash, and C. Yang, "Intensify the hindrance to gas sentrapment on the construction of Al 5356 thin-walled structure by tuning the WAAM process parameters," *Physica Scripta*, Oct. 2024, 99 (12), 125901, doi: 10.1088/1402-4896/ad896e
3. Q. Gao, F. Lyu, L. Wang, and X. Zhan, "Investigation of arc stability in wire arc additive manufacturing of 2319 aluminum alloy," *Materials*, Oct. 2024,*17*(21), 5160, doi: 10.3390/ma17215160
4. P. Dai et al., "Research status and development trend of wire arc additive manufacturing technology for aluminum alloys," *Coatings*, Aug. 2024, 14(9), 1094, doi: 10.3390/coatings14091094
5. R. Krein and V. Sushko, "Wire arc additive manufacturing of creep strength enhanced ferritic steels and nickel alloys," *In Advances in Materials Technology for Power Plants,* Oct. 2024, 84871, 495-506, doi: 10.31399/asm.cp.am-epri-2024p0495
6. H. He et al., "Open-source software architecture for multi-robot wire arc additive manufacturing (WAAM)," *Applications in Engineering Science*, 22, 100204, Aug. 2024, doi: 10.48550/arxiv.2408.04677
7. P. Chauhan, V. Vaghasia, R. Chaudhari, and J. Vora, "Experimental investigations on the fabrication of low alloy steels using wire arc additive

method," *The Eurasia Proceedings of Science, Technology, Engineering & Mathematics*, Aug. 2024, 28, 484-491, doi: 10.55549/epstem.1523859

8. K. Saravanakumar, V. G. Balaji, T. Srijha, V. Sanjay, and K. Thatchuneswaran, "Analysis of microstructure and mechanical properties of INCONEL 625 alloy by wire arc additive manufacturing (WAAM)," *Archives of Metallurgy and Materials*, Sep. 2024, doi: 10.24425/amm.2024.150928
9. Q. Shen, J. Xue, Z. Zheng, X. Yu, N. Ou, and J. Li, "Microstructures and mechanical properties of Co-based Elgiloy fabricated by wire arc additive manufacturing," *Science and Technology of Welding and Joining*, Sep. 2024, doi: 10.1177/13621718241280804
10. M. Harman, C. Çetînkaya, O. Yılmaz, and N. Bol, "Comparative process parameter optimization for wire arc additive manufacturing (WAAM) of E120C-GH4 metal cored and ER120S-G solid wire," *Politeknik Dergisi*, Aug. 2024, doi: 10.2339/politeknik.1478172
11. J. Franke, F. Heinrich, and R. Reisch, "Vision based process monitoring in wire arc additive manufacturing (WAAM)," *Journal of Intelligent Manufacturing*, Mar. 2024, doi: 10.1007/s10845-023-02287-x
12. D. Baier, T. Weckenmann, S. Baehr, and M. F. Zaeh, "Approach towards a quality assurance system for wire and arc additive manufacturing," *Processes*, Feb. 2023, doi: 10.3390/pr11020612
13. F. Coutinho et al., "iWAAM: an automated system for monitoring and control of wire-arc additive manufacturing," *Advances in Control and Optimization of Dynamical Systems*, Nov. 2023, 56(2), 6576–6581, doi: 10.1016/j.ifacol.2023.10.309
14. V. Carl, "Monitoring system for the quality assessment in additive manufacturing," *In AIP Conference Proceedings,* Apr. 2015, 1650(1), 171-176, doi: 10.1063/1.4914607
15. H. Kim, Y. Lin, and T.-L. B. Tseng, "A review on quality control in additive manufacturing," *Rapid Prototyping Journal*, Apr. 2018, *24*(3), 645-669, doi: 10.1108/RPJ-03-2017-0048
16. B. M. Colosimo, "Quality monitoring and control in additive manufacturing," In *Wiley StatsRef: Statistics Reference Online* , May 2020, doi: 10.1002/9781118445112.STAT08241
17. Y. Al-Meslemi, K. Ferreira, N. Anwer, A.-F. Obaton, H. Nouira, and N. Anwer, "Quality control for additive manufacturing," in: *Springer Handbook of Additive Manufacturing*, Jan. 2023, Springer, doi: 10.1007/978-3-031-20752-5_47
18. M. Rashid, S. Sabu, A. Kunjachan, M. Agilan, T. Anjilivelil, and J. D. R. Joseph, "Advances in wire-arc additive manufacturing of nickel-based superalloys: heat sources, DfAM principles, material evaluation, process parameters, defect management, corrosion evaluation and post-processing techniques," *International Journal of Lightweight Materials and Manufacture*, Jun. 2024, doi: 10.1016/j.ijlmm.2024.05.009
19. N. V. Prajna, "Wire arc additive manufacturing," *Mechanik, Werkstoffe und Konstruktion im Bauwesen*, Jan. 2023, doi: 10.1007/978-3-658-41540-2_4
20. M. S. M. Mansor et al., "Integrated approach to wire arc additive manufacturing (WAAM) optimization: harnessing the synergy of process parameters

and deposition strategies," *Journal of Materials Research and Technology*, May 2024, doi: 10.1016/j.jmrt.2024.03.170

21. J. P. Kumar, R. A. Prakash, and R. Raman, "Wire ARC additive manufacturing of functional metals—a review," *International Journal of Research and Review*, Jun. 2023, doi: 10.52403/ijrr.20230671
22. T. K. Goswami, S. Biswas, S. Das, and M. Mukherjee, "An overview of wire arc additive manufacturing (WAAM) technique with different alloys in modern manufacturing industries," *Hybrid Metal Additive Manufacturing*, Oct. 2023, doi: 10.1201/9781003406488-2
23. I. Khan, S. Anjimoon, V. Asha, A. Singla, P. Akhil, and N. M. A. Alsultany, "Advancing wire arc additive manufacturing: a new era of precision and efficiency in large metal component production," *2024 OPJU International Technology Conference (OTCON) on Smart Computing for Innovation and Advancement in Industry 4.0*, Jun. 2024, doi: 10.1109/otcon60325.2024.10687893
24. A. Nagpal, V. Alekhya, B. Swathi, A. Sravani, A. Kumar, and M. R. Shleghm, "Transforming wire arc additive manufacturing: a novel approach to achieving high deposition rates with reduced costs,"*2024 OPJU International Technology Conference (OTCON) on Smart Computing for Innovation and Advancement in Industry 4.0*, Jun. 2024, doi: 10.1109/otcon60325.2024.10688335
25. S. Chauhan and A. Narasimhulu, "Comprehensive study on wire arc additive manufacturing (WAAM)," *International Conference on Advances in Mechanical Engineering and Material Science*. Singapore: Springer Nature Singapore, Oct. 2023, doi: 10.1007/978-981-99-5613-5_22
26. M. Hietala, T. Rautio, M. Keskitalo, M. Jaskari, J. Arvola, and A. Järvenpää, "Impact of porosity on the bending fatigue strength of WAAM-fabricated super duplex stainless steel," *2024 International Conference on Power, Energy and Innovations (ICPEI). IEE*, Oct. 2024, doi: 10.1109/icpei61831.2024.10748952
27. J. Wei, C. He, R. Dong, N. Tian, and G. Qin, "Enhancing mechanical properties and defects elimination in 2024 aluminum alloy through interlayer friction stir processing in wire arc additive manufacturing," *Materials Science and Engineering A—Structural Materials Properties Microstructure and Processing*, Apr. 2024, *901*, 146582, doi: 10.1016/j.msea.2024.146582
28. B. Tomar, S. Shiva, and T. Nath, "A review on wire arc additive manufacturing: processing parameters, defects, quality improvement and recent advances," *Materials Today Communications*, May 2022, *31*, 103739, doi: 10.1016/j.mtcomm.2022.103739
29. X. Chen et al., "An in-process multi-feature data fusion nondestructive testing approach for wire arc additive manufacturing," *Rapid Prototyping Journal*, Oct. 2021, doi: 10.1108/RPJ-02-2021-0034
30. M. Liu et al., "Detection and quantitative evaluation of surface defects in wire and arc additive manufacturing based on 3D point cloud," *Virtual and Physical Prototyping*, Dec. 2023, doi: 10.1080/17452759.2023.2294336
31. J. Z. Wang, K. Zhu, W. Zhang, X. Zhu, and X. Lu, "Microstructural and defect evolution during WAAM resulting in mechanical property differences

for AA5356 component," *Journal of Materials Research and Technology*, Dec. 2022, doi: 10.1016/j.jmrt.2022.11.116
32. X. Chen et al., "A review on wire-arc additive manufacturing: typical defects, detection approaches, and multisensor data fusion-based model," *The International Journal of Advanced Manufacturing Technology*, Aug. 2021, doi: 10.1007/S00170-021-07807-8
33. H. Dai et al., "Wire arc additive manufacturing of ZL205A: heat input effect on the forming quality, pore defects and mechanical properties," *Journal of Alloys and Compounds*, Nov. 2024, doi: 10.1016/j.jallcom.2024.175777
34. A. Nagpal, A. Rajitha, K. Aravinda, G. Gouthami, R. Kalra, and N. R. Abdulla, "Breaking barriers in wire arc additive manufacturing: innovative solutions for enhanced geometrical accuracy and surface quality," In *2024 OPJU International Technology Conference (OTCON) on Smart Computing for Innovation and Advancement in Industry 4.0,* Jun. 2024, doi: 10.1109/otcon60325.2024.10688110
35. M. A. Machado, J. P. Oliveira, and T. G. Santos, "Non-destructive testing inspection for metal components produced using wire and arc additive manufacturing," *Superalloys*, Mar. 2023, doi: 10.3390/met13040648
36. R. Zimermann et al., "Multi-layer ultrasonic imaging of as-built wire + arc additive manufactured components," *Additive Manufacturing*, Dec. 2021, doi: 10.1016/J.ADDMA.2021.102398
37. A. Gardahaut, A. Charau, F. L. Bourdais, A. Stolidi, and N. Sargeeva-Chollet, "Implementation of non destructive testing and in-line monitoring techniques on extra-large structures printed with WAAM technology," *E-Journal of Nondestructive Testing*, Sep. 2023, doi: 10.58286/28477
38. P. Biswas, A. Rajitha, V. Revathi, H. P. Thethi, S. H. Mohammed, and D. K. Yadav, "Revolutionizing wire arc additive manufacturing: advances in geometric accuracy and surface finish optimization," *In 2024 OPJU International Technology Conference (OTCON) on Smart Computing for Innovation and Advancement in Industry 4.0*, Jun. 2024, doi: 10.1109/otcon60325.2024.10687444
39. F. H. Abdullah, K. Aswini, M. Manjunatha, H. P. Thethi, A. Parmar, and B. G. Bhavani, "Redefining the landscape of wire arc additive manufacturing: pioneering innovations for residual stress mitigation and process efficiency," *In 2024 OPJU International Technology Conference (OTCON) on Smart Computing for Innovation and Advancement in Industry,* Jun. 2024, doi: 10.1109/otcon60325.2024.10688195
40. T. Stepinski, "Ultrasonic nondestructive inspection of solid objects," *AIP Conference Proceedings,* Jan. 2011.
41. A. Ramos and J. L. S. Emeterio, "Ultrasonic systems for non-destructive testing using piezoelectric transducers: electrical responses and main schemes," *Piezoelectric Transducers and Applications. Berlin, Heidelberg: Springer Berlin Heidelberg*, Jan. 2009, doi: 10.1007/978-3-540-77508-9_16
42. B. Schneider, M. R. Shoaib, H. Taheri, L. W. Koester, T. A. Bigelow, and L. J. Bond, "Nondestructive evaluation of additively manufactured metal components with an eddy current technique," *ASNT 27th Research Symposium Proceedings,* Mar. 2018, doi: 10.32548/RS.2018.019

43. J. L. Rose, R. M. German, and D. D. Hongerholt, "Advances in ultrasonic inspection methods for PIM process monitoring and control," *Nondestructive Characterization of Materials VIII. Springer, Boston, MA,* Jan. 1998, doi: 10.1007/978-1-4615-4847-8_55
44. H. Mu, F. He, L. Yu, P. A. Commins, H. Wang, and Z. Pan, "Toward a smart wire arc additive manufacturing system: a review on current developments and a framework of digital twin," *Journal of Manufacturing Systems*, Apr. 2023, doi: 10.1016/j.jmsy.2023.01.012
45. N. Brown and S. Simske, "Detecting defects in sequential inputs to digital twins using machine learning," *IEEE Open Journal of Systems Engineering*, Jan. 2024, doi: 10.1109/ojse.2024.3458841
46. X. Zhu and Y. Ji, "A digital twin-driven method for online quality control in process industry," *The International Journal of Advanced Manufacturing Technology,* Jun. 2021,119(5), 3045-3064, doi: 10.21203/RS.3.RS-571586/V1
47. F. Dettù, B. Lakshminarayanan, S. Formentin, and C. R. Rojas, "From data to control: a two-stage simulation-based approach," *2024 European Control Conference (ECC). IEEE,* Jun. 2024, doi: 10.23919/ecc64448.2024.10591185
48. T. Abass, E. O. Itua, T. Bature, and M. A. Eruaga, "Concept paper: innovative approaches to food quality control: AI and machine learning for predictive analysis," *World Journal of Advanced Research and Reviews*, Mar. 2024, doi: 10.30574/wjarr.2024.21.3.0719
49. G. Mattera, J. Polden, and J. Norrish, "Monitoring the gas metal arc additive manufacturing process using unsupervised machine learning," *Welding in The World*, Sep. 2024, doi: 10.1007/s40194-024-01836-z
50. G. Mattera, "Defect monitoring in wire arc additive manufacturing using frequency domain analysis," *ESAFORM 2024,* May 2024, 41, 50, doi: 10.21741/9781644903131-6
51. H. Mu, F. He, P. Commins, D. Ding, and Z. Pan, "A digital shadow approach for enhancing process monitoring in wire arc additive manufacturing using sensor fusion," *Journal of Industrial Information Integration*, Apr. 2024, doi: 10.1016/j.jii.2024.100609
52. S. Oueslati, "Multi-sensor in process monitoring for WAAM: detection of process instability in electrical signals," *Mater Res Proceed,* May 2024, *41*, 371-379. doi: 10.21741/9781644903131-42
53. G. Nijhawan, L. Gummidi, V. Asha, V. Suresh, P. Praveen, and Z. H. Abdulzahraa, "The future of wire arc additive manufacturing comprehensive strategies for improved close loop monitoring and control," *2024 OPJU International Technology Conference (OTCON) on Smart Computing for Innovation and Advancement in Industry 4.0. IEEE,* Jun. 2024, doi: 10.1109/otcon60325.2024.10687734.
54. "Machine learning in process monitoring and control for wire-arc additive manufacturing," *Transactions on Intelligent Welding Manufacturing*, Dec. 2022, doi: 10.1007/978-981-19-6149-6_2
55. S. Karuthapandi, R. Murugan, and P.K. Palani, "Study and analysis of the macrostructure characteristics in FCAW with the use of a flat wire electrode

and by optimizing the process parameter using the Taguchi method and regression analysis," *High Temperature Material Processes*, 2016, 20(3).
56. S. Karuthapandi, M. Ramu, and P.R. Thyla"Effects of the use of a flat wire electrode in gas metal arc welding and fuzzy logic model for the prediction of weldment shape profile," *Journal of Mechanical Science and Technology*, 2017, 31(5).
57. B. M. A. Hamed, "Defect detection and process monitoring for wire arc additive manufacturing using machine learning," *Transactions on Intelligent Welding Manufacturing*, Jan. 2022, doi: 10.1007/978-981-19-3902-0_1
58. S. Kokare et al., "Wire arc additive manufacturing of a high-strength low-alloy steel part: environmental impacts, costs, and mechanical properties," *The International Journal of Advanced Manufacturing Technology*, Jul. 2024, 134(1), 453-475, doi: 10.1007/s00170-024-14144-z
59. Y. Li, C. Su, and J. M. Zhu, "Comprehensive review of wire arc additive manufacturing: hardware system, physical process, monitoring, property characterization, application and future prospects," *Results in Engineering*, Mar. 2022, *13*, 100330, doi: 10.1016/j.rineng.2021.100330
60. L. Jienan, Y. Xu, G. Yu, Z. Hou, and S. Chen, "Wire and arc additive manufacturing of metal components: a review of recent research developments," *The International Journal of Advanced Manufacturing Technology*, Sep. 2020, *111*(1), 149-198, doi: 10.1007/S00170-020-05966-8

Chapter 12

Revolutionizing production

The smart factory of tomorrow in additive manufacturing

G. Kalusuraman, K. Karthigadevi, G. Srinivasagan, and K. Sripriyan

12.1 INTRODUCTION

The rise of Industry 4.0 has revolutionized traditional manufacturing, incorporating advanced technologies that enhance the efficiency and flexibility of production processes. Additive manufacturing (AM), known for its ability to produce complex, customized parts, is at the forefront of this transformation. However, to fully capitalize on the potential of AM, it is crucial to optimize and manage the entire production process in an integrated and efficient manner. This is where the concept of the Virtual Smart Factory (VSF) comes into play, combining digital technologies such as the Internet of Things (IoT), artificial intelligence (AI), cloud computing, and virtual simulations to create a fully connected, adaptive manufacturing environment. A VSF allows for real-time data exchange between machines, sensors, and software, enabling manufacturers to make informed decisions that improve overall performance. By implementing key technologies such as digital twins, predictive maintenance, automated process control, and virtual simulations, the VSF offers a proactive approach to optimizing AM processes. It provides the ability to simulate and adjust production methods before they are executed in the real world, thus accelerating prototyping, lowering costs, and ensuring higher quality. Furthermore, AI-powered analytics continuously monitor and adapt the production environment, fostering a more sustainable and flexible manufacturing process. Ultimately, the VSF represents a significant leap forward in manufacturing technology, offering greater scalability, customization, and innovation in an increasingly competitive global market.

Manufacturing industries play a vital role in the economic development and growth of a country. As a result, continuous advancements in technology aim to refine manufacturing processes, boost product yield, and enhance quality. One such breakthrough technology that has recently gained significant attention is "smart manufacturing" (SM) [1, 2]. This approach enables precise prediction of product requirements and rapid identification of errors, leading to improvements in manufacturing processes and the

DOI: 10.1201/9781003569633-12

innovation of products and services. Currently, SM is crucial in increasing output with minimal human intervention while also enhancing quality, sustainability, and reducing costs.

At the heart of SM is the IoT, which facilitates seamless communication between machines and products. It allows for the integration of customer-specific instructions directly into the manufacturing process [3]. This system is referred to as a cyber–physical system (CPS), where computer algorithms monitor the manufacturing process by coordinating physical elements with computational systems [4]. Smart devices and sensors play a key role in this system, and advancements in sensing technologies, along with reduced sensor costs, have made this approach more feasible. The use of these sensors in industrial applications is known as the Industrial Internet of Things (IIoT) [5], which enables efficient decision-making through the analysis of collected data.

SM involves the full integration of manufacturing processes, using data analytics, modeling, simulation, and real-time decision-making to address changing demands in the factory, supply network, and customer requirements [6]. The ultimate goal is to produce customized, eco-friendly, and intelligent products and services [7]. The ongoing industrial transformation, often referred to as Industry 4.0, leverages CPS to automate and streamline manufacturing operations while sharing real-time information. Key technologies such as IIoT, machine learning (ML), AI, and Big Data analytics are enhancing data mining and industrial control, contributing to the development of "smart factories" [7]. Factories today are governed by real-time interactions between machines, people, and the entire production system. A large amount of data is gathered from various sources, including production processes, product defects, customer feedback, market demands, and analytics. This data is crucial for making informed decisions to optimize production efficiency and product yield. Furthermore, cloud computing plays a key role by enabling networked data storage and management, facilitating off-site analysis, and providing users with easy access to results via mobile devices [7].

To ensure the successful operation of smart factories, three core objectives must be met: (1) collecting a wide range of relevant data from equipment, (2) analyzing the data to generate actionable insights for both operational and business decision-making, and (3) leveraging automation and system optimization to inform business strategies [8]. A smart factory's components can be divided into three categories: (1) SM, which focuses on the seamless integration of tools, machines, and operators; (2) digital services, which involve real-time monitoring of manufacturing and maintenance, as well as forecast analysis; and (3) smart machines, which enhance the performance of manufacturing execution systems (MES) or shop floors, waste reduction efforts, and material requirements planning (MRP),

As per the report by the McKinsey Global Institute, approximately 60% of all occupations involve at least 30% of tasks which could be automated, highlighting the potential for SM technologies to significantly increase the production capabilities of smart factories [9]. SM has applications across various industries, including food, pharmaceuticals, semiconductors, oil and gas, chemicals, paints, and additives [8]. This growth is driven by increased investments in automation technologies, which are expected to further propel the SM sector.

12.2 EVOLUTION OF SM

Manufacturing has evolved drastically over the years, from steam- and water-powered processes in the 18th century to the sophisticated SM techniques of today. The current era is defined by the fourth industrial revolution, which follows a long history of innovations and continuous improvements in manufacturing methods. The timeline of the four manufacturing revolutions is depicted in Figure 12.1.

12.2.1 First industrial revolution

The first industrial revolution began with the development of the modern steam engine by Thomas Newcomen in the 1700s, which had a transformative effect on manufacturing processes. In the 1760s, the United Kingdom introduced steam- and water-powered machines for mass production, which spread to Europe and America between 1760 and 1840 [10]. The textile industry was one of the first to adopt these new manufacturing methods, attracting significant investments [11]. Other sectors, including iron production, mining, and agriculture, also benefitted from these innovations. Key developments during this period included the development of coal and iron, the creation of railroads, and the mass manufacturing of textiles. The widespread use of coal and the steam engine created a highly efficient energy source, while the development of railroads and an influx of investments further accelerated manufacturing processes.

Despite its advancements, the first industrial revolution also faced significant challenges, including poor working conditions, child labor, environmental pollution, and unhealthy living conditions for workers [12].

12.2.2 Second industrial revolution

The second industrial revolution, which took place between 1870 and 1914, brought about further technological advancements that elevated manufacturing processes to new heights. This era saw the widespread use of new energy sources, including electricity, gas, and oil, facilitated by the development of more advanced combustion engines. The rapid expansion of

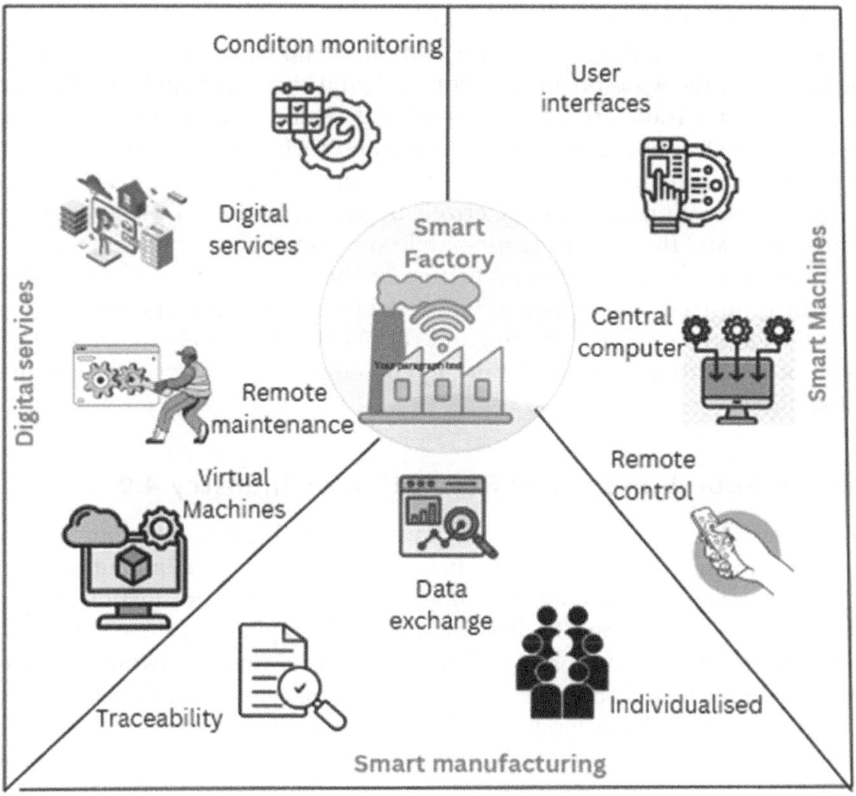

Figure 12.1 Smart factory.

railroads and telegraph lines after 1870 played a key role in accelerating the exchange of ideas, transportation of goods and manpower, and the movement of manufacturing equipment. This period marked the onset of globalization, which significantly eased the production process and led to an increase in production output.

The advent of electricity, combined with innovations in telecommunication such as the telephone, helped propel the production sector to a more innovative level through factory electrification and the introduction of the production line. Electric power became a standard form of energy transmission, which enables the creation of vacuum pumps, gas lighting systems, electric generators, and transformers. The steel industry grew rapidly due to the increased demand for steel, while new chemical synthesis techniques brought about the development of synthetic fabrics, dyes, and fertilizers. The arrival of automobiles and commercial airplanes further revolutionized transportation at the turn of the 20th century [11].

12.2.3 Third industrial revolution

The third industrial revolution, also known as the digital revolution, occurred after the significant disruptions of World War I and II. This period, beginning in 1969, saw the widespread adoption of digital computers and data-keeping systems, which transformed mechanical and analog technologies into digital systems. Nuclear energy emerged as a new, powerful energy source, marking a significant leap from previous energy sources. This phase also escorted in the information age, characterized by the growth of microprocessors, transistors, and the increasing use of telecommunications and computers.

Automation became a central feature of this era, driven by the introduction of industrial robots and programmable logic controllers (PLCs) [13]. The development of the Internet and the World Wide Web accelerated the pace of innovation in manufacturing, enabling more advanced production capabilities.

12.2.4 Fourth Industrial Revolution or Industry 4.0

The Fourth Industrial Revolution, or Industry 4.0 (I4.0), was first introduced by Germany in 2011, marking the beginning of a new era in manufacturing. The concept centers around the computerization of the production process through the integration of CPSs. Industry 4.0 emphasizes the connectivity provided by the IoT, ML, and real-time data processing, allowing machines to communicate and share real-time information with one another.

Industry 4.0 is characterized by the union of digital and physical systems, which enables intelligent decision-making in the manufacturing process. This era has seen the widespread adoption of advanced technologies such as AI, flexible automation systems augmented reality (AR), robots, and AM. Key features of I4.0 include (1) full automation in manufacturing processes, (2) integration of IIoT- and CPS-based systems, (3) optimization of production steps based on client demand, (4) closed-loop data models and control systems, and (5) the ability to order customized products [14]. The main goal of Industry 4.0 is to enable autonomous decision-making within manufacturing, allowing for more efficient and adaptable production methods. As industries adopt these technologies, the vision of Industry 4.0 is gradually being realized, refer Figure 12.2–12.5.

12.3 CURRENT STATE-OF-THE-ART IN SM

Since the introduction of Industry 4.0 in Germany in 2011, many other regions around the world have developed strategies to integrate SM into their industries. After Germany's lead in adopting CPS-based manufacturing processes, countries like the United States, Japan, China, and Taiwan have actively focused on technological innovations to embrace this shift.

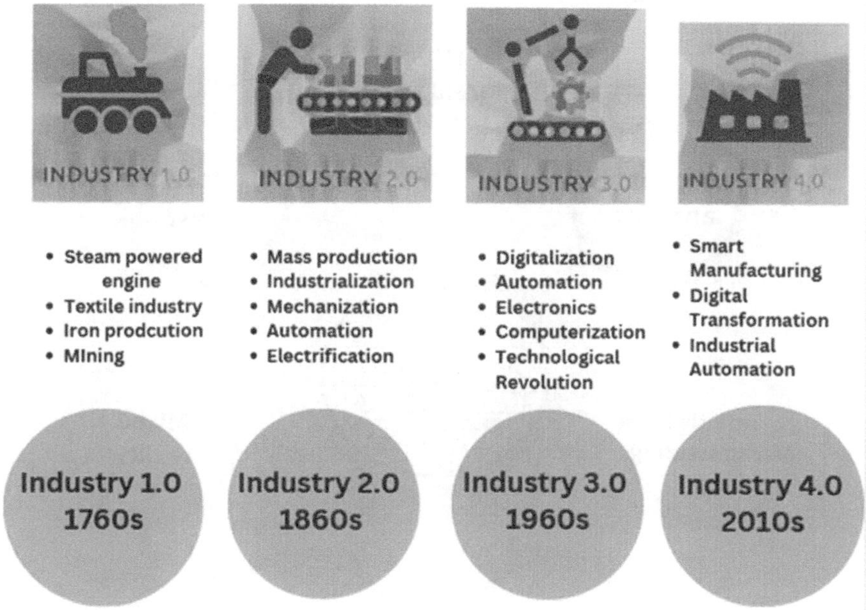

Figure 12.2 The industries revolutions – Timeline and defining characteristics.

12.3.1 The United States

In 2012, General Electric (GE) introduced the concept of the IIoT, suggesting that the future of manufacturing lies in integrating smart equipment, advanced production systems, and effective decision-making tools. To accelerate the adoption of IIoT, the Industrial Internet Consortium (IIC) was established in 2014 with the collaboration of companies such as GE, Cisco, AT&T, Intel, and IBM. The IIC aims to provide resources, pilot projects, and expertise related to IIoT technologies, including their security.

The United States has emphasized the role of information technologies (ITs), such as virtual reality (VR), Big Data, and cloud computing, as essential for implementing IIoT. Recently, the IIC has partnered with CESMII, a non-profit institute dedicated to SM in the United States, to accelerate the digital transformation of manufacturing processes. These collaborations help enhance the adoption of IIoT technologies, offering manufacturers the benefits of increased visibility on the shop floor, improved supply chain management, reduced operational costs, and energy consumption optimization. Furthermore, these technologies support proactive equipment maintenance, improved sustainability, and new business opportunities, ultimately boosting profitability and cost savings for manufacturers [13].

Figure 12.3 Technologies driving smart manufacturing.

12.3.2 Japan

In response to Germany's Industry 4.0, Japan initiated the Industrial Value Chain Initiative (IVI) in 2015 to integrate manufacturing and IT and build a connected ecosystem. This initiative involves top industries in Japan, focusing on distinguishing collaborative areas between companies. By organizing these areas and connecting participating companies' technologies, Japan has fostered the development of a highly interconnected manufacturing system.

The IVI's activities include business scenario groups focused on using CPSs to enhance manufacturing processes. Additionally, Japan has set up platform workgroups to facilitate the development of platforms for production planning, supply chain management, quality control, and asset management. To ensure security, Japan also introduced the "Industry 4.1 J" program, which emphasizes a higher level of security in SM systems, including real-time data analysis and cloud-based data accumulation.

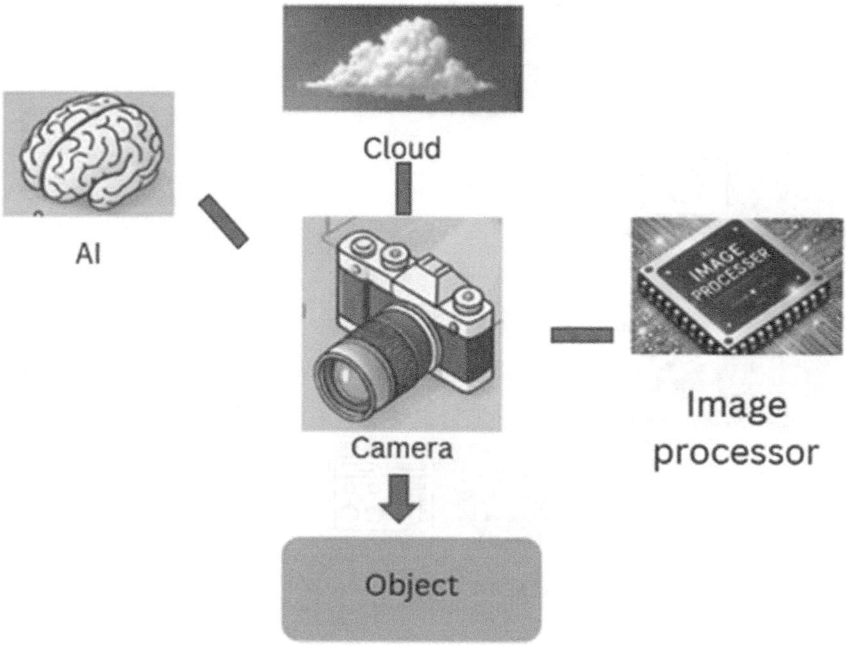

Figure 12.4 A standard machine vision system for product quality inspection.

12.3.3 China

In 2015, China launched the "Made in China 2025" (MIC25) initiative, a 10-year program aimed at achieving technological self-reliance and advancing its manufacturing capabilities. The initiative focuses on improving innovation, merging industrialization with IT, enhancing manufacturing quality, promoting environmentally sustainable practices, and fostering international collaborations.

China's first step in the MIC25 program was the promotion of cloud manufacturing, which serves as the foundation for SM. The country has already implemented advanced systems in specific areas, including computerized numerical control (CNC) machines, industrial robots, and AM. To support the widespread adoption of SM, China's Ministry of Industry and Information Technology (MIIT) funded numerous pilot projects, which span various manufacturing categories such as discrete manufacturing, process manufacturing, and mass-customized manufacturing [15].

One example of a successful implementation of SM in China is Haier, a leading manufacturer of household appliances. Haier has adopted the concept of "connected factories," where customers, products, machines,

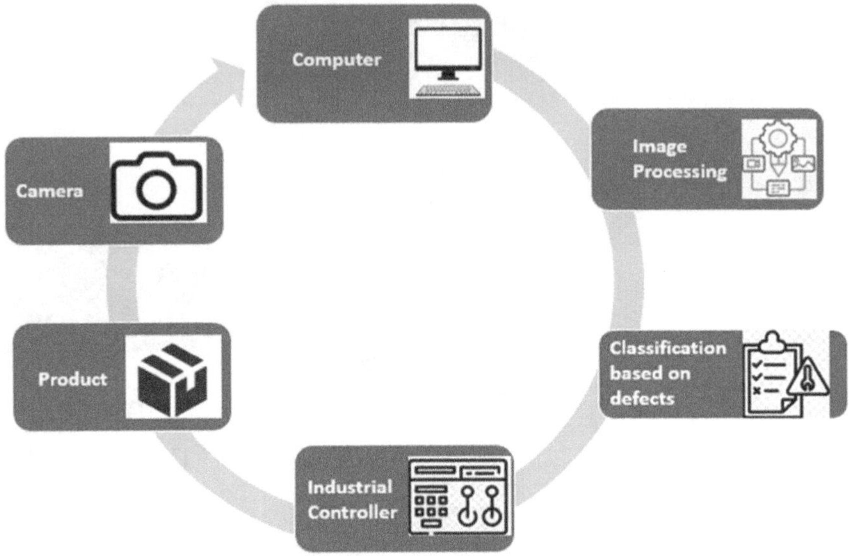

Figure 12.5 Schematic diagram – AI.

and factories are interconnected. This approach has significantly reduced product delivery times and operational costs, showcasing the potential of SM in improving efficiency and business outcomes.

12.3.4 Taiwan

Taiwan, known for its dominance in semiconductor chip manufacturing, has actively embraced Industry 4.0 through various SM initiatives. In 2015, Taiwan launched the "Productivity 4.0" plan to enhance its industrial technologies and establish itself as a key player in the global manufacturing supply chain.

Taiwan also developed the "five plus two" innovative industries plan in 2016, which uses Industry 4.0 technologies such as cloud computing, the IoT, Big Data, and smart robotics to enhance the country's manufacturing capabilities. As the fourth largest machine tool exporter in the world, Taiwan has great potential to establish smart factories.

12.4 TECHNOLOGIES ASSOCIATED WITH SM

SM relies on a wide array of technologies that help optimize and streamline manufacturing processes. These technologies include immersive technologies such as VR, AR, and mixed reality (MR), all of which significantly enhance the manufacturing experience. Below, we explore these technologies in more detail.

12.4.1 Immersive technologies

Immersive technologies, which blend sight, sound, and touch to create an enriched digital experience, have revolutionized the way industries interact with applications, content, and experiences. In the situation of SM, immersive technologies play a significant role in enhancing design, training, and real-time decision-making. The main types of immersive technologies are VR, AR, and MR. Here, contributions to SM are discussed.

12.4.1.1 Virtual reality

VR has seen significant implementation in manufacturing sectors, particularly in product design and simulation. VR technology allows manufacturers to visualize product designs in the early stages without physical prototypes, reducing prototyping time and costs. It also enables the visualization of products in simulated environments, which helps improve customization options and design modifications.

Additionally, VR is used in automated guided vehicles (AGVs) and future factory (FoF) initiatives, allowing for three-dimensional (3D) visualization of AGV movement within factory simulations. VR-based setups, such as motion-capturing devices combined with smartwatches, also facilitate human–robot collaboration (HRC) by enabling virtual commissioning of systems. Furthermore, VR technology is applied to training young engineers by providing detailed, virtual-based education in industrial processes, mechanization, and maintenance systems, contributing to enhanced learning in SM environments [16, 17].

12.4.1.2 Augmented reality

AR blends the digital and physical worlds by overlaying digital content, such as images, sounds, and videos, on real-world objects. This technology improves human–machine interaction, enabling operators to assess and respond to manufacturing tasks more effectively. AR provides real-time information, which enhances decision-making efficiency, quality, and overall productivity.

In manufacturing, AR can assist with logistics analysis, product design reviews, and real-time maintenance and assembly instructions. AR technology will not require complex, exclusive hardware and can often be implemented using standard equipment. It also facilitates training, allowing new technicians to learn how to service or repair equipment by visualizing the disassembly and internal components without prior hands-on experience.

AR has been employed successfully in various manufacturing scenarios. Both VR and AR technologies enhance the manufacturing process by speeding up production line support, improving shop floor operations, and enabling more efficient training, diagnostics, and warehouse management [12].

12.4.1.3 Mixed reality

MR combines the physical and virtual worlds to create an environment where digital and real-world objects coexist and interact in real time. MR devices are increasingly being used in SM systems. They are particularly valuable in training maintenance staff, as they can simulate the machinery environment, allowing trainees to acquire skills and experience before interacting with real-world equipment. By providing schematics and interactive models of machinery, MR helps technicians easily modify or repair faulty components.

In addition to training, MR is beneficial for providing immersive visual representations of machine integration, system functionalities, and data streams that drive SM. MR technology also enhances product sales by offering virtual demonstrations of products, service instructions, maintenance guidance, and seamless knowledge transfer to end-users.

12.4.2 Additive manufacturing

AM, or 3D printing, plays a significant role in SM by enabling rapid, precise, and flexible production of complex components. This technology is especially useful in research and development (R&D), where manufacturing limitations often prevent quick prototype fabrication. With AM, prototypes can be created quickly using accurate materials and design specifications, significantly reducing development time and costs [18].

AM rushes the design process and supports mass customization, allowing for quick modifications based on market request. The AM process can be controlled online remotely, allowing factories to integrate multiple 3D printers to create user-specific products. This flexibility makes it easier for companies to optimize their products and scale up production efficiently [18].

Moreover, AM enables small-scale companies and individual users to design and manufacture their own prototypes or products. Customers can customize product designs, materials, colors, and aesthetics, while also checking the existing materials and machines, making the process more personalized. Common materials used in AM include polymers, metals, alloys, and ceramics [14].

12.4.3 Big Data Analytics

Big Data analytics is a crucial technique in SM, allowing for the collection and analysis of vast amounts of data generated by production units, and product request systems and customer feedback. This data analysis aids in real-time decision-making, optimizing production processes and improving efficiency [10].

The use of Big Data enables several benefits: (1) the generation of knowledge from continually changing data, (2) tracking value streams to monitor irregularities, (3) enhancing model optimization for predictive outcomes, (4) identifying new diagnostic possibilities, and (5) optimizing key performance indicators (KPIs) [19].

For example, GE's factory in New York uses advanced sensors to gather data on manufacturing processes and building conditions. These insights are evaluated in real time, helping employees quickly identify issues and enhance the quality and efficiency of the production process. Similarly, companies like Intel and BMW have leveraged Big Data analytics to reduce testing costs, predict failures, and enhance product quality [19]. Big Data thus empowers manufacturers to make better, data-driven decisions and predict failures before they occur.

12.4.4 Industrial Internet of Things

IIoT is a foundational technology in Industry 4.0. IIoT involves embedding sensors throughout production lines to monitor equipment performance and process outputs. These sensors are connected via cloud technology, facilitating the integration of all production steps into a seamless system [19].

IIoT allows for better production planning by enabling smart process control, predictive maintenance, fault detection, and more efficient human–machine interactions [19]. The data collected from IIoT sensors is stored in the cloud, and through ML and Big Data analytics, manufacturers can identify and address issues before they lead to costly breakdowns.

Additionally, IIoT transforms traditional factories into smart factories by enabling interconnected digital supply networks (DSNs). These networks make factories more energy-efficient, cost-effective, and resilient to operational risks [1]. By optimizing quality control, supply chain traceability, and equipment fault prediction, IIoT helps manufacturers save time and reduce production costs while improving overall factory performance.

12.4.5 Artificial intelligence

AI plays a pivotal role in optimizing the use of robots and minimizing human participation in high-risk areas. AI systems are capable of making autonomous decisions, optimizing processes, responding to changes in production schedules, addressing machine malfunctions, automatically replacing parts, and triggering alarms in case of critical issues [20]. It also contributes to enhancing factory performance, controlling product costs, optimizing processes, and improving efficiency.

Through predictive and prescriptive analytics powered by ML, AI significantly improves the efficiency of assembly workers, offering greater predictability and certainty in operations. AI learns patterns in production facilities with HRC, solving complex problems. For example, in scenarios requiring precise control over temperatures, pressures, and liquid flow, AI-based systems can manage these factors more [21] accurately and safely than humans. AI also reduces downtime in continuous production lines, which can otherwise severely affect product yields.

The rapid development of AI has enabled the integration of digital twin technology into manufacturing. Digital twins involve the virtual replication of physical systems, allowing for computational simulations that help identify design flaws early in the process. This reduces the need for costly reconfigurations of manufacturing systems The configuration, motion, control, and optimization (CMCO) framework is a model used for designing flow-type SM systems. Digital twins facilitate hardware-in-the-loop simulations that prevent early-stage design errors, and the integration of AI algorithms helps dynamically optimize system operations based on real-time data [22].

AI in Industry 4.0 leverages deep learning to process data, recognize speech, reconstruct images, interpret natural language, and more. Deep learning systems learn from large volumes of data, identifying patterns and making decisions that improve manufacturing processes. Various types of analytics—descriptive, predictive, diagnostic, and prescriptive—are applied to enhance decision-making. These innovative analytics transform manufacturing into highly optimized smart systems [23–25].

ML plays a crucial role in SM, supporting tasks like pattern recognition, fault detection, quality control, and monitoring. ML algorithms are typically divided into supervised learning, unsupervised learning, and reinforcement learning, all aimed at improving classification, regression, and other tasks in manufacturing [26]. For instance, ML models such as artificial neural networks (ANN) are commonly used in cutting processes like milling and turning. ML enhances productivity and product quality by detecting defects and predicting maintenance needs before equipment failures occur [26].

As industries move toward full automation, data management and processing become even more critical, and ML algorithms are increasingly leveraged to handle these complex tasks [27].

12.5 INFLUENCE OF AI IN SM

12.5.1 Quality checks

Quality inspections are a vital part of the manufacturing process, ensuring that products meet required standards before reaching the market. However, traditional manual inspections are often time-consuming and prone to human error, resulting in inconsistent product quality. This is where AI-based systems, such as machine vision technology, become crucial.

Machine vision combines high-resolution cameras, AI, and edge computing to automate the inspection of product quality. These AI-driven systems can detect defects in products [28], packaging, and labeling, and make decisions based on the observed conditions. Unlike standard machine vision systems, which can notice errors but not classify them, AI-based machine vision can not only spot issues but also decide on corrective actions This advancement helps maintain consistent product quality and improves overall efficiency in production, leading to higher customer satisfaction and loyalty.

12.5.2 Machine vision-based automated optical inspection

Automated optical inspection (AOI), driven by machine vision technology, is essential in manufacturing, as it helps enhance product quality, increase productivity, and boost competitiveness. A typical AI-powered machine vision setup consists of a smart camera with necessary sensors and AI-integrated software. The camera is linked to cloud storage, where like images are stored. The system is capable of adjusting its working distance automatically, a mechanism known as pre-processing automation. Once an image is captured, it undergoes various transformations like converting color images to grayscale and standardizing them to uniform dimensions. Advanced algorithms then analyze the extracted image data to make predictions, and defective products are rejected based on these analyses.

12.5.3 Maintenance

AI-powered sensors on smart factory floors collect vast amounts of data, which is analyzed in real time to assist decision-making and reduce production downtime. Predictive maintenance is one such application, where Prognostics and Health Management (PHM) systems identify potential faults in machines before they occur [29]. These systems combine data science, physics, and statistics to evaluate machine conditions and predict when maintenance is necessary, minimizing unscheduled downtime. The PHM process involves three main steps: (1) *observe*—data collection and processing, (2) *analyze*—condition assessment, diagnostics, and prognosis, and (3) *act*—human–machine interactions for decision-making and action.

12.5.4 Reliable design

In modern SM, AI is used for generative design, which explores thousands of potential designs by considering various input parameters such as dimensional constraints, material choices, and mechanical strength. AI-based software generates innovative design solutions that would be difficult for humans to conceive, providing a significant number of design options. This approach reduces material use, product weight, and cost while maintaining strength and performance.

These designs are often validated through AM, which creates prototypes and allows for iterative refinement before moving to mass production. By integrating IoT, customer feedback can be incorporated into future design iterations, making the process even more dynamic and responsive.

12.5.5 Reduced environmental impact

Manufacturing is responsible for a significant portion of global energy consumption, contributing to around 50% of total energy usage, with a large amount of waste stemming from low-quality products and excess raw materials. AI can significantly reduce the environmental footprint of manufacturing by optimizing resource use and reducing waste. It helps minimize the overuse of harmful materials, optimizes the generation of scrap waste, and balances energy distribution across manufacturing processes. By combining AI with technologies like generative design and AM, manufacturers can create products with less waste and environmental impact compared to traditional methods [30]. AI also improves the ability to monitor and manage production processes, further reducing inefficiencies and energy usage.

12.5.6 Factory integration

AI, IIoT, and IoT are increasingly being integrated into smart factories to streamline operations from material selection to product delivery. In addition to optimizing internal production processes, AI enables the integration of multiple manufacturing plants, facilitating the sharing of data across factories. This interconnectivity allows for automated material provisioning, predictive maintenance, and other efficiencies that save both time and resources. By optimizing the flow of data and resources, factory integration supports a more agile, efficient, and responsive manufacturing environment.

12.5.7 Post-production support

AI also plays a significant role in post-production support by helping manufacturers provide clients with real-time updates about product performance. AI can predict when failures might occur or when maintenance is necessary.

12.6 AUTOMATED OPTICAL INSPECTION IN SM

In the era of Industry 4.0, ensuring high-quality control is critical for any manufacturing process. AOI systems are widely used because they provide a non-destructive method of inspecting products before they reach the market, helping prevent supply chain disruptions and maintain high standards of quality. AOI systems are particularly important for their ability to assess products with great precision, relying on advanced imaging technologies that exceed the capabilities of the human eye. While humans can only detect a limited range of electromagnetic wavelengths (390–770 nm), AOI cameras can be customized to capture a broader spectrum of wavelengths, enabling more detailed and accurate inspections.

12.6.1 Conventional AOI system

The conventional AOI system has evolved significantly with advancements in image capture and processing technologies. This development has led to improvements in the precision of AOI-based quality inspections, making it a vital tool in the manufacturing sector. Unlike manual inspections, AOI systems use these advanced technologies to provide faster, more consistent, and more accurate inspections, making them an essential component of quality control in the Industry 4.0 landscape.

The hardware configuration of AOI systems typically includes image sensors and specialized lighting arrangements to capture digital images of products. On the software side, various inspection algorithms are employed to analyze these images and detect defects. Recently, the integration of 3D imaging technology utilizing fringe projection has been introduced to enhance inspection capabilities. This technology enables the detection of more complex issues, such as identifying lifted leads in surface-mounted devices (SMDs) or verifying the coplanarity of ball grid arrays (BGAs), which are often challenging for conventional top-down cameras to identify. Moreover, advancements in Industry 4.0 technologies have enabled AOI systems not only to produce but also to store vast amounts of data, enhancing their inspection capabilities. Currently, AOI systems are primarily used for inspecting printed circuit boards (PCBs) to detect defects such as nodules, scratches, stains, open circuits, and solder joint thinning. These systems are also used to inspect semiconductor wafers, light-emitting diodes (LEDs), and flat-panel displays (FPDs), covering a range of inspection tasks, including surface-level defects, image analysis, pattern matching, color verification, and product dimensions [31–33].

12.6.2 AI-powered AOI

As electronic device designs have grown more intricate, traditional AOI systems have struggled with accuracy during inspections. AI-based AOI systems, however, learn from large datasets of both good and defective products, allowing the software to build models that can efficiently identify specific defects. The more the AI system scans products, the more accurate it becomes. AI-powered high-resolution cameras perform real-time quality checks at any point in the production line, enabling early detection of defects and helping manufacturers identify the origin of issues during manufacturing stages. This integration of AI with AOI systems significantly improves the speed and accuracy of inspections and reduces the need for manual operations. Manufacturers that have adopted AI-based AOI systems have reported much higher accuracy rates, with some achieving 97% accuracy compared to 60–70% for conventional systems.

For instance, AI-based 3D-AOI has greatly reduced false alarms in solder defect inspections, which are typically challenging for traditional systems. The AI system's ability to detect these defects in a smart factory environment greatly enhances production quality. An example of this is seen in an ML model used to assess the pattern integrity of inkjet-printed electrical devices, where the model evaluates line edge roughness (LER) and its impact on the electrical properties of the devices. This kind of model, integrated into AOI systems, helps optimize smart factory production lines for high precision and accuracy [34–36].

12.7 CHALLENGES IN SM

Despite the benefits, SM faces significant challenges. As manufacturing technologies become more advanced, the systems themselves are more complex and integrated, which introduces new security concerns. The growing reliance on CPSs, with their interconnected devices, increases vulnerability to cyberattacks, and many companies remain ill-prepared for these risks. These challenges are particularly pronounced for small- and medium-sized enterprises (SMEs), which struggle to address such issues due to limited resources [37–40].

12.7.1 Data security

Data security is one of the most pressing concerns for smart factories. With the extensive use of IIoT, IoT, and cloud computing to interconnect operational technology (OT), IT, and intellectual property (IP), there is an increased risk of cyberattacks. Data breaches can lead to severe disruptions, including the loss of real-time data monitoring, predictive maintenance capabilities, and supply chain management. A study found that 65% of companies view IoT technologies as a major source of cybersecurity risks. Cyberattacks can bring smart factory operations to a standstill, making robust cybersecurity measures essential.

12.7.2 System integration and interoperability

As smart factories adopt cutting-edge technologies, compatibility issues between new and existing machinery can arise. One of the greatest challenges in SM is ensuring seamless machine-to-machine (M2M) communication. For instance, while modern manufacturing systems excel under internet protocol version 6 connectivity, some older devices may not be compatible with newer systems. This lack of interoperability can create significant challenges in achieving the full potential of SM.

12.7.3 Safe HRC

Collaborative robots (cobots) are central to Industry 4.0 and play a significant role in enhancing factory efficiency by reducing human errors. These robots are designed to work alongside humans safely in a shared workspace, improving human–machine interactions and streamlining manufacturing processes. There are concerns regarding the safety of on-site personnel, particularly if the cobots malfunction or if workers are not adequately trained to collaborate with them. Additionally, mobile cobots, which are battery-powered, can face downtime if their charge is depleted during tasks. To address this, cobots can be programmed to automatically return to their charging stations when their battery levels drop below a predefined threshold.

Another challenge is that cobots often struggle to differentiate between different raw materials, which may limit their efficiency in certain contexts. Furthermore, while cobots can work autonomously to an extent, they still require human supervision to ensure their proper functioning, particularly for quality checks at specific intervals. This reliance on human oversight can be burdensome for employees, leading to fatigue.

12.8 FUTURE PROSPECTS OF SM IN THE GLOBAL MARKET

The COVID-19 pandemic has accelerated the demand for remote working capabilities and improved supply chain agility, both of which can be facilitated by SM technologies Consequently, there is a growing demand for employees with the technical skills necessary to manage smart factories. The most sought-after skills include effective communication, teamwork, flexibility, analytical thinking, and knowledge of smart technologies.

SMEs are expected to increasingly adopt SM technologies. To assist SMEs in this transition, a structured framework can be followed, which includes: (1) identifying available manufacturing data, (2) assessing the readiness of the SME for SM, (3) setting tailored objectives based on the SME's needs, and (4) selecting the appropriate tools and practices for achieving those objectives. By adopting such a framework, SMEs can cultivate a data-driven decision-making culture.

While many advanced factories are incorporating elements of SM, key characteristics such as self-optimization, self-configuration, predictive maintenance, and autonomous decision-making are still lacking in many facilities

[16]. The adoption of 5G technology and edge computing is expected to accelerate the integration of cloud manufacturing and Big Data analytics, further enhancing the capabilities of smart factories. Additionally, with the IoT, end-users will be able to directly request services tailored to their specific needs, driving further advancements in the field.

12.9 CONCLUSION

SM represents a crucial component of the fourth industrial revolution, offering the potential for tailored solutions, enhanced product quality, and improved productivity across factories. This article has explored the evolution of industrial manufacturing technologies and provided an overview of the current state of SM as a key element of Industry 4.0. We have also highlighted the role of AI in revolutionizing manufacturing processes and examined various SM technologies, including the challenges faced by the industry today. These challenges provide an avenue for future research and innovation for both academics and industry practitioners.

Manufacturers are progressively embracing full automation to minimize labor costs and boost production efficiency. AI-based quality inspection technologies are crucial in ensuring that only high-quality products are delivered to the market. Furthermore, innovative concepts like immersive technologies, AM, the IIoT, and CPSs are significantly transforming manufacturing processes, driven by intelligent systems.

However, these advanced technologies are still not widespread in many manufacturing sectors, leading to a significant gap between the ideal SM system and the traditional manufacturing systems in place. This gap may be attributed to the challenges associated with integrating new technologies and the uncertainties around their return on investment. Although the initial adoption of SM can be challenging, the long-term benefits for enterprises and end-users are substantial, providing a strong incentive for continued innovation and adoption.

REFERENCES

1. Kusiak A (2018) Smart manufacturing. *Int J Prod Res* 56:508–517
2. Maheswari MG, Karthigadevi K, Rani GE (2024) A novel breast cancer detection system from mammographic images using a hyperparameter tuned gated recurrent unit with attention included capsnet. *J Biomol Struct Dyn* 1–14
3. Yang H, Kumara S, Bukkapatnam STS, Tsung F (2019) The internet of things for smart manufacturing: a review. *IISE Trans* 51:1190–1216
4. Lee EA (2015) The past, present and future of cyber–physical systems: a focus on models. *Sensors* 15:4837–4869

5. Sripriyan K, Ramu M, Thyla PT (2023) Characterization of dissimilar welding of AISI 409 ferritic to AISI 316l austenitic stainless steel using AISI 308 filler wire under different temperature conditions. In: *Advances in Additive Manufacturing and Metal Joining: Proceedings of AIMTDR 2021*, Springer. pp. 309–321
6. Munirathinam S (2020) Industry 4.0: industrial internet of things (IIOT). In: *Advances in Computers*. Elsevier, pp. 129–164
7. Lu Y, Xu X, Wang L (2020) Smart manufacturing process and system automation—a critical review of the standards and envisioned scenarios. *J Manuf Syst* 56:312–325
8. Manyika J, Chui M, Miremadi M, et al. (2017) A future that works: AI, automation, employment, and productivity. *McKinsey Glob Inst Res Tech Rep* 60:1–135
9. Moghaddam M, Cadavid MN, Kenley CR, Deshmukh A V (2018) Reference architectures for smart manufacturing: a critical review. *J Manuf Syst* 49:215–225
10. Sahoo S, Lo C-Y (2022) Smart manufacturing powered by recent technological advancements: a review. *J Manuf Syst* 64:236–250
11. Phuyal S, Bista D, Bista R (2020) Challenges, opportunities and future directions of smart manufacturing: a state of art review. *Sustain Futur* 2:100023
12. Abidi M, El-Tamimi A, Al-Ahmari A (2012) Virtual reality: next generation tool for distance education. *Int J Adv Sci Eng Technol* 2:95–100
13. Zhang C, Yang J (2020) *A History of Mechanical Engineering*. Springer
14. Mehrpouya M, Dehghanghadikolaei A, Fotovvati B, et al. (2019) The potential of additive manufacturing in the smart factory Industrial 4.0: a review. *Appl Sci* 9:3865
15. Zhou K, Liu T, Zhou L (2015) Industry 4.0: towards future industrial opportunities and challenges. In: *2015 12th International Conference on Fuzzy Systems and Knowledge Discovery (FSKD)*. IEEE, pp. 2147–2152
16. Zhong RY, Xu X, Klotz E, Newman ST (2017) Intelligent manufacturing in the context of industry 4.0: a review. *Engineering* 3:616–630
17. Taxén G, Naeve A (2002) A system for exploring open issues in VR-based education. *Comput Graph* 26:593–598
18. Damiani L, Revetria R, Volpe A (2015) Augmented reality and simulation over distributed platforms to support workers. In: *2015 Winter Simulation Conference (WSC)*. IEEE, pp. 3214–3215
19. Ren L, Zhang L, Wang L, et al. (2017) Cloud manufacturing: key characteristics and applications. *Int J Comput Integr Manuf* 30:501–515
20. Nagorny K, Lima-Monteiro P, Barata J, Colombo AW (2017) Big Data analysis in smart manufacturing: a review. *Int J Commun Netw Syst Sci* 10:31–58
21. Evans PC, Annunziata M (2012) Industrial internet: pushing the boundaries. *Gen Electr Reports* 488–508.
22. Serrano-Ruiz JC, Mula J, Poler R (2021) Smart manufacturing scheduling: a literature review. *J Manuf Syst* 61:265–287

23. Liu Q, Leng J, Yan D, et al. (2021) Digital twin-based designing of the configuration, motion, control, and optimization model of a flow-type smart manufacturing system. *J Manuf Syst* 58:52–64
24. Wang J, Ma Y, Zhang L, et al. (2018) Deep learning for smart manufacturing: methods and applications. *J Manuf Syst* 48:144–156
25. Teti R, Jemielniak K, O'Donnell G, Dornfeld D (2010) Advanced monitoring of machining operations. *CIRP Ann* 59:717–739
26. Wu D, Jennings C, Terpenny J, et al. (2017) A comparative study on machine learning algorithms for smart manufacturing: tool wear prediction using random forests. *J Manuf Sci Eng* 139:71018
27. Du Preez A, Oosthuizen GA (2019) Machine learning in cutting processes as enabler for smart sustainable manufacturing. *Procedia Manuf* 33:810–817
28. Sharp M, Ronay Ak, Hedberg Jr T (2018) A survey of the advancing use and development of machine learning in smart manufacturing. *J Manuf Syst* 48:170–179
29. Wang W-C, Chen S-L, Chen L-B, Chang W-J (2016) A machine vision based automatic optical inspection system for measuring drilling quality of printed circuit boards. *IEEE Access* 5:10817–10833
30. Sripri K, Murugan R, Thyla PR, Anantharuban K (2022) Characteristic of weld bead using flat wire electrode in GMAW inline during the process: an experimental and numerical analysis. *Int J Press Vessel Pip* 196(January).
31. Biggio L, Kastanis I (2020) Prognostics and health management of industrial assets: current progress and road ahead. *Front Artif Intell* 3:578613
32. Hernandez Korner ME, Lambán MP, Albajez JA, et al. (2020) Systematic literature review: integration of additive manufacturing and Industry 4.0. *Metals (Basel)* 10:1061
33. Abd Al Rahman M, Mousavi A (2020) A review and analysis of automatic optical inspection and quality monitoring methods in electronics industry. *IEEE Access* 8:183192–183271
34. Janóczki M, Becker Á, Jakab L, et al. (2013) Automatic optical inspection of soldering. *Mater Sci Top* 2156–3950
35. Richter J, Streitferdt D, Rozova E (2017) On the development of intelligent optical inspections. In: *2017 IEEE 7th Annual Computing and Communication Workshop and Conference (CCWC)*. IEEE, pp. 1–6
36. Hsu S-M, Lo C-Y (2019) Advanced qualification method for patterns with irregular edges in printed electronics. *Flex Print Electron* 4:15001
37. Liu T-J, Hsu S-M, Wu M-J, et al. (2019) Efficient and improved qualification method for patterns with irregular edges in printed electronics. *J Micromech Microeng* 29:124005
38. Liu T-J, Wu M-J, Lo C-Y (2020) Machine learning-based off-line electrical characteristic prediction through in-line pattern integrity inspection. *J Micromech Microeng* 31:15005
39. Tuptuk N, Hailes S (2018) Security of smart manufacturing systems. *J Manuf Syst* 47:93–106
40. Mittal S, Khan MA, Romero D, Wuest T (2018) A critical review of smart manufacturing & Industry 4.0 maturity models: implications for small and medium-sized enterprises (SMEs). *J Manuf Syst* 49:194–214

Index

For Product Safety Concerns and Information please contact our EU representative GPSR@taylorandfrancis.com Taylor & Francis Verlag GmbH, Kaufingerstraße 24, 80331 München, Germany